METAL CONTAMINATION
OF FOOD

CONOR REILLY

B.Sc., B.Phil., Ph.D.

*Head of Department of Public Health and Nutrition,
Queensland Institute of Technology, Brisbane, Australia*

APPLIED SCIENCE PUBLISHERS LTD
LONDON

APPLIED SCIENCE PUBLISHERS LTD
RIPPLE ROAD, BARKING, ESSEX, ENGLAND

British Library Cataloguing in Publication Data

Reilly, Conor
  Metal contamination of food.
  1. Food contamination   2. Metals—Toxicology
  I. Title
  614.3′1   TX571.M/

ISBN 0–85334–905–3

© APPLIED SCIENCE PUBLISHERS LTD 1980

Photoset by Thomson Press (India) Ltd , New Delhi
Printed in Great Britain by Galliard (Printers) Ltd, Great Yarmouth

METAL CONTAMINATION OF FOOD

Preface

Some eighty of the hundred-odd elements of the Periodic Table are metals. Several of the metals are known to be essential for human life; others are toxic, even in small amounts. These metals occur in all foodstuffs, in greater or lesser amounts depending on various circumstances. The presence of metals in food can have both good and bad consequences. It can be of interest to food processors, nutritionists, toxicologists and a wide variety of other scientists. This book is concerned entirely with the presence of these metals in food. It concentrates on metals which are generally considered undesirable, at least in more than trace quantities, in food. In a general way it follows the pathways by which those metals get into food, and examines the significance—from the point of view of the manufacturer as well as of the scientist—of that contamination. It considers how man has sought to protect himself from the undesirable effects of metal contamination of food by passing laws and regulations. The international consequences of such laws are evidenced by the efforts of the United Nations and other organisations to develop a uniform and harmonised universal code of regulations concerning food.

In the second part of the study attention is given to individual metals which occur as food contaminants. A great deal of investigation has been carried out on many of these metals from biochemical, medical and other points of view. The findings of these studies have been published in many different journals and some are still only available in restricted form. An attempt has been made in this book to present a brief and useful summary of available information gathered from many sources. The aim has been to make available in one place what otherwise might require time-consuming literature searches.

Metals have served man well since he started on the long road of technological progress. We chart human progress by the metals we have used: thus we refer to the Bronze and Iron Ages. Only relatively recently have we begun to appreciate the significance of metals in a facet of human life besides that of technology. We now know a great deal about the part played by metals in the structure and function of the human body. We classify some metals among the essential nutrients. We know that others are destructive of human life and development. But we are also aware that certain metals can have both good and bad effects on the human organism, depending on the amount of metal present and other factors. Apart from the biological and medical scientist, the food chemist and technologist also have a major interest in the metals. They know the importance of metals in food not only with regard to health and toxicity, but also from the point of view of food quality, processing and commercial characteristics.

The general public is more aware today than ever before of the part metals in food play in people's lives. They are exposed to advertising extolling the nutrient value of minerals in food and in health supplements of many kinds. They are alert to the dangers of heavy metal contamination of what they eat and look to legislation to protect themselves against an excess of undesirable metals in their diet.

It is with these diverse groups of people in mind that this book was written. They all have a commendable interest in the topic, but information has not always been readily available to meet their needs. True, the specialist will frequently want to delve deeply into the question and seek out original reports, but often the intelligent layman, as well as the technician and specialist, will want a comprehensive, general treatment of the subject such as is given here.

Metal contamination of food is by no means confined to any one nation. Data given here on metals in Israeli canned food, Australian fish, Polish margarine and Zambian gin show the international nature of the problem. Water supplies in Boston as well as in Glasgow, as will be seen, can carry the same contaminating metals as do illicit spirits in Kentucky. For this reason summaries of food laws relating to metals from the UK, the US, Australia and other English-speaking countries, as well as the current international codes, are given in the text. It is hoped that, in this way, the basic data on metals in food and their significance for human health will be related to the practical context of the day-to-day user and the food producer.

Gratitude is owed to many for help given in writing this work: to my wife Ann, especially, for encouragement and practical help with references and bibliography; to Drs George and Olga Berg of the University of Rochester, New York, for setting me right on US food legislation, and for other advice and help; to my brother Brian who, from the point of view of his own profession of engineering, contributed to sections on metal contamination during food processing; and to many others who will remain anonymous. Finally, a word of gratitude to the hard-working secretaries who deciphered my writing and endured my re-editing with so much patience.

CONOR REILLY
Brisbane, 4 March, 1980

Contents

PART II. THE INDIVIDUAL METALS

PART I

General

1

'A Peck of Dirt'

'Every man must eat a peck of dirt before he dies', says an old English proverb, and, in spite of all the pure food laws that have been promulgated throughout the world during the past century, it is difficult to find fault with this piece of folk wisdom. Whether he likes it or not, a man will, during his lifetime, consume a large quantity of matter which, while we might not want to call it dirt, is certainly not nutrient. If we were to judge solely by the lists of contents of many books on food and nutrition, we might be led to conclude that all that we get on our plates at mealtime are the proteins, carbohydrates, fats, inorganic nutrients and vitamins, with some less well-defined substances contributing to flavour and colour. Yet analyses of the raw materials used by the cook in preparing our meals would show that the true picture was far from that painted by the neat tables of the *Manual of Nutrition*.[1] Apart from an extraordinary array of organic molecules, almost any foodstuff will contain most of the 80 or so metals and metalloids of the Periodic Table of the elements. Only a few of these are nutrients strictly so-called; many, not a few of which are present in relatively large quantities, might be described as nutritionally indifferent; others are potentially toxic. These metals form part of the 'dirt' of the proverb and by many might be described as contaminants.

Contamination is a difficult concept to define accurately. We all know, more or less, what it means. As the *Concise Oxford Dictionary* tells us, to contaminate means to pollute or infect. Hence, a contaminant is something that pollutes. Pollute, in its turn, the dictionary tells us, means to destroy the purity or sanctity of something, to make it foul or filthy. Scientists use the term contaminant in a more precise manner to describe 'undesirable materials which have been added inadvertently before, during or after processing of food'.[2] From the point of view of the food processor the term

3

contaminant will imply what must be avoided in the products he sells to the public, including such impurities as traces of arsenic, lead and other toxic metals, so as to comply with prescribed legal standards or, in their absence, with accepted codes of practice. The consumer will think of contamination as the addition to his food of anything which would make it unfit or injurious to health and not of the quality he has the right to expect.

When considering the metals from the point of view of food contamination some special problems arise which are not, for instance, encountered to anything like the same extent in the case of contamination by micro-organisms or mycotoxins. Though several metals are normally described as toxic even in very small amounts and others are toxic when they exceed a certain level which is often not precisely known, in fact most metals will be found occurring naturally in almost any food sample examined. Indeed, two of the very toxic substances of this class, selenium and cobalt, are not only present in many foodstuffs but are included among the essential trace nutrients. No doubt other metals which are today believed to be highly undesirable will, in future years, be accepted as essential nutrients and so the problem of knowing what is and what is not a metal contaminant will become even more difficult. It is for such reasons that most countries have specialist committees with an ongoing brief to monitor the most recent findings on the status of metals in food and to recommend new or amended legislation whenever this becomes necessary. The tasks that face such committees are, as this volume will help to show, considerable, for the question of metals in food and their relation to human health is complex and as yet far from resolved.

THE METALS

Some 80 of the 104 elements now listed in the Periodic Table are metals. Another 17 can be described as non-metals, leaving a small intermediate group of what are called metalloids. The distinctive physical and chemical properties of the metals and the non-metals are summarised in Table 1.[3] The differences in chemical properties between the two groups are mainly related to the fact that atoms of non-metals can readily fill their valence shells by sharing electrons with or transferring electrons from other atoms. Metallic character decreases and non-metallic character increases with an increase in the number of valence electrons. In addition, metallic character increases with the number of electron shells. Thus the properties of succeeding elements change gradually in progressing across the Periodic

TABLE 1
PROPERTIES OF METALS AND NON-METALS

Metals	Non-metals
Physical properties	*Physical properties*
1. Good conductors of heat and electricity	1. Poor conductors
2. Malleable and ductile in solid state	2. Brittle, non-ductile in solid state
3. Metallic lustre	3. No metallic lustre
4. Opaque	4. Transparent or translucent
5. High density	5. Low density
6. Solids (except mercury)	6. Gases, liquids, or solids
7. Crystal structure in which each atom is surrounded by eight to twelve near neighbours; metallic bonds between atoms	7. Molecules consist of atoms covalently bonded; the noble gases are monatomic
Chemical properties	*Chemical properties*
1. One to four electrons in outer shell; usually not more than three	1. Usually four to eight electrons in outer shell
2. Low ionisation potentials; readily form cations by losing electrons	2. High electron affinities; readily form anions by gaining electrons (except noble gases)
3. Good reducing agents	3. Good oxidising agents (except noble gases)
4. Hydroxides basic or amphoteric	4. Hydroxides acidic
5. Electropositive; oxidation states positive	5. Electronegative; oxidation states either positive or negative

Table, with an increase in the number of valence electrons from left to right and down the groups, when the number of electron shells increases. There is, consequently, no sharp distinction but rather a merging, between metals and non-metals. Elements on the borderline show some of the characteristics of metals and some of non-metals. This intermediate group is known as the metalloids. The group includes boron, silicon, germanium, arsenic, selenium, antimony and tellurium. Because of their metallic characteristics and their importance as contaminants in food, some of these will be included, without any particular distinction being made, in the present study.

METALS IN THE ENVIRONMENT

Analysis of the tissues of the human body will show the presence of most of the metallic elements, in greater or lesser amounts. This is not surprising since the food we eat also contains a wide variety of metals, reflecting the distribution of these elements in the environment. The soil in which plants grow contains metals from a number of sources, principally the rocks from which the soil was formed, as well as fertilisers, sewage sludge and other materials added in the course of agricultural activities. Metals are also contributed by the debris of mining and industrial waste, by the dust and smoke of fossil fuel combustion and by other forms of atmospheric pollution. Water too makes its contribution to an extent related to the source of supply and the degree of pollution.

The actual amount of metal found in any one soil sample will depend on the nature of the parent rocks, their degree of mineralisation and other factors. Table 2 shows the overall abundance in the lithosphere of a number of metals and the range of concentrations found in soils. The distribution in soil of some of the less common though still very widely distributed metals, which often occur in 'trace' or parts per million (mg/kg) amounts in food, are shown in Table 3.

The atmospheric distribution of metals and other elements often reflects industrial activity rather than geological structures, although locality, such as proximity to sea, can result in elevated levels of sodium in air. A report[5] of a study on 30 trace elements in the atmosphere carried out in the UK draws attention to the fact that even distant industrial activity can contribute to the metal content of the air in rural districts. Lead, for example, was 380 ng/kg of air in industrialised Nottinghamshire and 139 ng/kg at rural Lake Windermere. Similarly, mercury and zinc were 0.09 and 415 ng/kg respectively at Nottingham and 0.13 and 132 ng/kg

TABLE 2

AVERAGE ABUNDANCE OF SOME METALS IN THE EARTH'S CRUST

Metal	Lithosphere content (mg/kg dry wt)	Soil content (mg/kg)
Iron	50 000	7 000–55 000
Manganese	1 000	200–5 000
Chromium	200	5–3 000
Vanadium	150	20–25
Nickel	100	10–800
Zinc	80	10–300
Copper	70	2–100
Cobalt	40	1–5
Molybdenum	2	

at Lake Windermere. As the authors note, results of similar investigations in Japan and the USA show concentrations of many metals to be more than 10 times those found in the UK. Nevertheless, the atmosphere at Windermere in its turn had levels of several metals exceeding those found in rural Canadian sites.

Concentrations of metals in water can reflect nearby industrial activity as well as the composition of local rock and soil. In addition reticulated water may carry metallic contributions due to the composition of plumbing and containers. Table 4 shows the levels of some metals in domestic water

TABLE 3

AVERAGE TRACE METAL CONTENT OF SOIL

Metal	Soil content (mg/kg)
Antimony	6
Arsenic	6
Barium	500
Beryllium	6
Cadmium	0.06
Lead	10
Mercury	0.3
Scandium	7
Selenium	0.2
Silver	0.1
Tellurium	10.1
Tin	10
Vanadium	100

TABLE 4

METALS IN DOMESTIC WATER SUPPLIES (USA)[6]

Metal	Average concentration (μg/litre)
Cadmium	1.3
Chromium	2.3
Cobalt	2.2
Copper	134.5
Iron	166.5
Lead	13.1
Manganese	22.2
Nickel	4.8
Silver	0.8
Zinc	193.8

supplies in an American city. Similar results have been reported for other US cities and for some European water supplies.

METALS IN FOOD

The metal content of food, whether this be of animal or of plant origin, will depend on many factors, ranging from environmental conditions to methods of production and processing. Even in the same class of food, variations in the levels of metals may be considerable. Zinc in seafoods, for example, may range from less than 5 mg/kg to as much as 1000 mg/kg, while nickel in meat may vary from zero to 4.5 mg/kg.[4] Within a particular foodstuff, levels of metals may vary between parts. While, for example, wheat germ contains 7.4 mg/kg of copper, endosperm may have less than 2.0 mg/kg. An egg may have 35.5 mg/kg of zinc in its yolk but only 0.3 in the white. Food processing can also cause similar differences. While unpolished rice contains 0.16 mg/kg of chromium, this is reduced to 0.04 mg/kg with polishing. In the same way iron is reduced by almost 76 per cent, manganese by 86 per cent and cobalt by 89 per cent when wheat is converted into white flour.[7]

Differences in brand of processed food can result in considerable variations in the levels of metals obtained in the diet. A study of several brands of evaporated milk and infant formulae showed that levels of iron in the milk ranged from 0.66 to 1.88 mg/kg and in the formulae from 0.82 to 19.01 mg/kg. The variations between levels of other metals were not as great but were still significant.[8] While these findings are of considerable importance with regard to the health of infants, of less significance, but of considerable interest at least for a section of the population, are the results of another study[9] on quite a different type of beverage. This concerned levels of metal contaminants in 'moonshine' whisky. Because of the crude equipment used by illicit distillers, metal uptake by their products can be considerable. The consequences for the health of drinkers can be serious. However, revenue officers in the USA have in some cases been able to identify the source of illegal spirits by their trace metal 'fingerprint', no doubt an advantage to the Excise Department. Twenty-two different metals were detected in the samples analysed in the investigation. Antimony, for example, had a range of 0.01–38.9, cadmium 18.1–37.6, copper 0.5–23.3, iron 0.2–1.9 and zinc 0.1–143 mg/litre.

THE METAL CONTENT OF THE HUMAN BODY

It is not surprising, then, since the environment in which we live, the air we breathe, the water we drink and the food we eat contain such a wide variety of metals, that our bodies also should show a wide range in kinds and concentrations of these elements. Most of the metal in our bodies has come in with our food. However, not all the metal we ingest is retained. Some is lost by excretion in faeces, urine and sweat, in hair and skin. The amount of metal actually absorbed by our body from food will depend to some extent on our choice of diet but also on our state of health and our genetic make-up, as well as factors such as vitamins in our food. In spite of such variability it is possible to draw up a table of average quantities, indicating dietary intake as well as levels of loss, which gives a useful picture of the status of metals in human nutrition, as is shown in Table 5. The total body levels of some of the metals are also included.

TABLE 5

INTAKE AND EXCRETION OF TRACE METALS[10,11]

Metal	Diet (mg/day)	Urine (mg/day)	Sweat (mg/day)	Hair (µg/g)
Essential				
Chromium	0.05–0.1	0.008	0.059	0.69–0.96
Manganese	2.2–8.8	0.225	0.097	1.0
Iron	15	0.25	0.5	130
Cobalt	0.3	0.26	0.017	0.17–0.28
Copper	3.2	0.06	1.59	16–56
Zinc	8–15	0.5	5.08	167–172
Selenium	0.068	0.04	0.34	0.3–13
Molybdenum	0.3	0.15	0.061	
Nickel	0.4	0.011	0.083	
Possibly essential				−
Vanadium	2.0	0.015	−	
Toxic				0.0075
Cadmium	0.215	0.03	−	−
Lead	0.450	0.03	0.256	2.8–4.8
Mercury	0.02	0.015	0.0009	18–19
Antimony	0.15	0.07	0.011	6
Beryllium	0.013	0.0013	−	6.5
Arsenic	1.0	0.195	−	−
Barium	1.25	0.023	0.085	2
Non-toxic				5
Tin	4.0	0.023	2.23	−
Rubidium	1.5	1.1	0.05	−
Aluminium	4.5	0.1	6.13	5
Titanium	0.85	0.33	0.001	0.05
Zirconium	4.2	0.14	−	−
Boron	1.3	1.0	−	7

Within the body itself the distribution of metals is by no means uniform. Some metals are accumulated in particular tissues and organs and others in different portions of the body. For example, in a 55 kg man, the toxic metal cadmium will be accumulated preferentially in kidney (12 mg) and to a much lower extent in blood (0.76 mg), whereas kidney will contain 0.12 mg and blood 1.3 mg of lead. Chromium is concentrated preferentially in muscle (2.4 mg) but only reaches 0.019 mg in the kidney.[11]

THE ROLE OF METALS IN THE HUMAN BODY

Among the many metals found in the body only a small number are known to be essential for normal life. The absence of these metals will result in the appearance of characteristic pathological deficiency symptoms. Most of the other metals present are artefacts, with no functional significance. These metal nutrients can be divided into two classes according to the amounts of each which are required for normal function. As is shown in Table 6, potassium, magnesium, calcium and sodium are classified as macro-nutrients and the remainder as micro-nutrients. The use of the term trace element for this second group is no longer as common as it was in the past. The term can cause some confusion, for some of the trace elements, so-called, are actually present in food and in the body in relatively significant amounts. It is the dietary requirement for them which is small. However, when the term is used it generally refers to inorganic components which occur at or below the milligram per kilogram range and can exert some influence on body metabolism. An examination of the literature on nutrition published over the past few decades will show how views on the composition of the second category of essential metals have changed. Twenty years ago only five or six of the micro-nutrients were considered to be essential for life, but since then, as investigations proceeded and as better methods of analysis were developed, several more metals have been recognised as being equally essential. It is fairly easy to demonstrate that a dietary insufficiency of a macro-nutrient will result in metabolic disorders, but it is not so for a micro-nutrient. For various reasons we cannot subject humans experimentally to deprivation of certain metals in the diet in order to see what symptoms may result. The experimental animals used for nutrition experiments have often such minute dietary requirements for some metals that it is almost impossible to produce a purified and balanced diet with a metal content sufficiently low to allow investigation of its essentiality. Even the environment of a metal cage and unavoidable traces in analytical grade chemicals can easily give rise to sufficient contamination to meet daily needs for a metal. Other factors may also cause complications. Thus establishing whether or not a metal is an essential micronutrient can be a tedious and sometimes inconclusive process. The list of metals in the micro-nutrient group given in Table 6 would, for such

TABLE 6

CLASSIFICATION OF METALS IN THE BODY

Grouping	Metal	Per cent body weight
1. Macro-nutrient metals essential for body function	Calcium	1.5–2.2
	Potassium	0.4
	Sodium	0.2
	Magnesium	0.05
2. Micro-nutrient metals essential for body function	Iron	0.003 5
	Zinc	0.002 5
	Selenium	0.000 3
	Manganese	0.000 2
	Copper	0.000 1
	Molybdenum	
	Cobalt	
	Chromium	
	Silicon	
	Nickel	
	Tin	
3. Metals for which essentiality has not yet been established, although there is evidence of their involvement in some cell reactions	Barium	
	Arsenic	
	Strontium	
	Cadmium	
	Vanadium	
4. Metals found in the body for which no metabolic function is known	Lead	
	Mercury	
	Gold	
	Silver	
	Bismuth	
	Antimony	
	Boron	
	Beryllium	
	Lithium	
	Gallium	
	Titanium	
	and others	

reasons, not be accepted by all nutritionists. Nickel and tin, for instance, would probably be moved by some into the third group of metals, and vanadium elevated to the second.

Besides the essential elements there are other metals which are widely distributed in small amounts in living tissue. Some of them appear to assist in vital activities but their absence does not cause any visible ill effects. These have been called 'beneficial elements' by some nutritionists. It is quite probable that these metals will eventually be shown to be essential, but requirements may be so minute as to be almost undetectable.

To some extent the known functions of the macro-nutrients and the micro-nutrients overlap. They all work in three main ways in the body: as constituents of bone and teeth; as soluble salts which help to control the composition of body fluids and cells; and as essential adjuncts to many enzymes and other functional proteins. The macro-nutrients play major roles in the first two functions while the micro-nutrients are especially prominent in assisting enzyme function. Very few of the proteins that act as biological catalysts can do so entirely on their own. Most need the assistance of a non-protein prosthetic group. If the prosthetic group is detachable it is known as a coenzyme. The group may be an organic molecule containing one of the trace metals, or it may consist solely of a trace element. In the latter case, if the metal is detachable from the protein part, it is known as an activator. Iron and copper, for example, occur in the prosthetic group of many enzymes concerned with oxidation–reduction reactions. Molybdenum has much the same function in other enzymes concerned with cellular oxidation. Zinc and manganese function as detachable activators on some of the enzymes involved in cellular metabolism. Most of the other micro-nutrients have been shown to play similar enzymatic roles, and as a consequence the enzymes involved are often referred to as metalloenzymes. The inorganic micro-nutrients are also found in some other body compounds including hormones and vitamins. The production and storage of insulin in the pancreas, for instance, involves zinc. Haemoglobin, essential for the transport of oxygen in the blood, is an iron-containing compound. Cobalt atoms form part of cobalamin or vitamin B_{12}. We shall look at some of these substances and the related functions of the metals in later sections. For the present it is sufficient to indicate briefly that those metals which are included in groups 1, 2 and 3 of Table 6 can be shown to play essential roles in human metabolism and body function. This cannot be said by any means of all the metals which occur in food.

THE TOXIC METALS

As will be made clear later, it is not always possible to draw a distinction between essential and toxic metals. All metals are probably toxic if ingested in sufficient amounts. Indeed, sometimes the margin between toxicity and deficiency is very small, for example in the case of selenium. In addition, it is difficult to consider toxicity of a single metal in isolation. Under normal conditions all metals interact in the body to varying degrees. The physiological effects, including the toxicity of cadmium, for example, are closely related to the amount of zinc also present. Likewise the function of iron in cells is affected by both copper and cobalt and to some extent also by molybdenum and zinc. Since several other similar interactions between the metals in the body are known, caution must be exercised when deciding whether or not a particular trace metal is toxic.

In spite of the above reservations it is possible to differentiate between elements which are known with certainty to be essential and those which display severe toxicological symptoms at extremely low levels and have no known beneficial function. Mercury, cadmium and lead qualify for inclusion in this group. They are among the most commonly encountered toxic metals in food. No doubt there are other metals, especially the radioactive elements such as the man-made plutonium, which also have no known beneficial function in the human body and are extremely hazardous even at low concentrations. These are the metals we will be considering in detail in this study of metal contamination of food. We will also give considerable attention to those micro-nutrients which become toxic when present in excess.

Perhaps it is a good idea here to consider briefly what we mean by the term 'toxic'. D. A. Phipps in his book *Metals and Metabolism*[12] defines a toxic metal as one which belongs to a 'group of elements which have neither an essential nor a beneficial but a positively catastrophic effect on normal metabolic functions, even when present in only small amounts'. What exactly these 'catastrophic effects' are in individual cases is not by any means always clear. Phipps warns that the pathological effects and significant dose in metal-ion poisoning are remarkably complex and variable, depending particularly on factors that can modify the uptake of the metal as well as those which control its subsequent metabolism. We shall see how variable the effects of toxic metals can be and how great is the difficulty of deciding when, for example, what is referred to as subclinical poisoning begins.

EFFECTS OF METALS IN FOOD

The concern of the food processor and supplier is not merely to see that his products are free of toxic metals or even of essential metals in quantities great enough to cause poisoning; nor is it entirely to meet the various legal requirements and observed agreed codes of practice. He also has to see to it that his products do not contain metals which might cause deterioration and quality defects. He must, moreover, be alert to the possibility of pick-up from non-food sources such as containers and piping of traces of metals which would affect the colour, texture or shelf-life of his foodstuffs: for trace metal contamination can result in all these defects.

Trace metals are responsible for many of the colour changes observed during the preparation and cooking of food. Even at levels of only a few milligrams per kilogram complexes can be formed between metal ions and plant pigments, causing the development of colours not all of which are welcome. Traditionally, green vegetables were cooked in unlined copper saucepans to bring about the development of a bright green pigment, which was considered to enhance the customer appeal of the food. However, copper at equally low levels will darken cherries and even turn them black. Iron, too, is responsible for several colour changes in food. It reacts with the anthocyanins in some fruit to produce a black pigment. It can impart a grey-green colour to cream and have a similar effect on certain chocolate-containing foods. Aluminium and tin can also cause darkening of food colours.

Another serious effect of trace metal contamination is the development of rancidity in fats. Traces of copper, iron and some other metals act as catalysts in the oxidation of unsaturated bonds in lipids and cause rapid and costly deterioration of cooking oils and fat-containing foods.

2

How Metals Get into Food

METALS IN SOIL

The soil is the main source of the metals we find in plants. All the nutrients a plant needs for growth, with the exception of carbon, are drawn directly from the soil. Soil is an exceedingly heterogeneous substrate. It consists of a solid phase composed of mineral matter, originally part of rocks of the lithosphere, and of organic matter which represents plant and animal debris in various stages of decomposition. It also has a liquid phase, the soil solution, containing dissolved salts, organic compounds and gases. There is, in addition, a gaseous phase, the soil atmosphere, occupying spaces between the solid particles which are not filled with water. Besides these three phases there is a microbial population living in the film of water which surrounds the soil particles and underground parts of plants. Bacteria, fungi and algae play a role in making the soil suitable for plant growth. In this they are assisted by the plants themselves, which secrete sugars and amino acids which serve as nutrients for the soil micro-organisms.

To grow crops satisfactorily soil must contain as a minimum an adequate supply of the macro- and micro-nutrients essential for plant growth. All soils will contain, in addition, many other elements, generally in very small amounts, in both organic and inorganic form; some of these elements will be beneficial, others indifferent and some potentially toxic to plants. There is great variation between the types and quantities of trace elements in different soils. This is true even where the soil is suitable for agricultural use and has not been contaminated by industrial debris or other sources of metal pollution.

Whatever the actual composition of the soil may be, the plant has to be able to absorb nutrients and carry them to where they are needed in its

15

tissues. The soil liquid phase is the most important immediate source of these nutrients. It is a very dilute solution and would be quickly depleted of its contents were it not for the fact that it is constantly replenished by release into it of elements from the solid phase. Metals are released from the solid phase reserves partly by solubilisation of soil minerals and organic matter and partly by ion exchange.

The needs of an actively growing plant could not normally be met by the nutrients supplied by the soil solution found in the immediate area below the stem. It is necessary for the plant to seek out adequate supplies of nutrients, even at some distance from its growing position. This it does by the development of roots as absorbing organs. These have evolved in such a way that they can penetrate far out into the soil, bringing the greatest possible absorbing surface into immediate contact with the soil matrix. An often quoted study of Dittmer[1] illustrated how extraordinary can be the extent of this contact between plant roots and soil, even in a modestly sized annual plant. Dittmer grew a single plant of rye, *Secale cereale*—a domesticated member of the grass family (Gramineae) —in a box of soil just under 56 cm deep and 30.5 × 30.5 cm in surface area. After four months of growth he carefully liberated the root system and measured its size. The total area of the root surface in contact with the soil was 639 m^2 and the combined length of all the roots was 523 km. Other studies have shown that the roots of a single plant of corn (*Zea mais*) can penetrate to a depth of more than 6 m and extend horizontally as much as 10 m within 14 weeks after planting the seed.[2] Undoubtedly, then, when the question of uptake of minerals from the soil by plants to be used for human food is being considered, the nature of the underlying subsoil—and not just the top few feet of cultivated land—is important.

ACCUMULATION OF METALS BY PLANTS

There is another complicating factor which should be taken into account when metal contamination of plant foodstuffs is being considered. While all plants will, if placed in a nutritionally balanced soil, take up nutrients to the extent they need for growth, others possess a special ability which enables them to take up and accumulate, sometimes to very high levels, certain specific elements. A well-known example of such behaviour is the accumulation of selenium by the legume *Astragalus racemosus*. Selenium is now recognised as an essential plant and animal nutrient. However, it is also highly toxic if present in more than minute amounts. Selenium is widely distributed in the earth's crust at concentrations of about 0.09 ppm and most vegetables and pasture plants may absorb about 5 ppm without signs of phytotoxicity. However, certain alkaline soils in parts of the USA and elsewhere have high levels of selenium. This prevents the growth of many fodder plants, but not of *Astragalus*, which grows well and accumulates remarkable quantities of the metal. Levels as high as 15 g/kg of plant tissue have been reported.[3] Animals which feed on prairies where *Astragalus* is prevalent suffer a severe type of poisoning known as alkaline disease.

Geobotanical Indicators

Plants such as *Astragalus* are often referred to as indicator plants. Since they grow on mineralised soils where other plants cannot survive and, in addition, often accumulate large amounts of particular metals in their tissues, they can be used as pointers to the presence of local mineral deposits. Many such plants are known. A species of basil, *Becium homblei*, known to prospectors in parts of Africa as the 'copper flower', has been used by them as an aid to prospecting.[4] Other indicator plants for a variety of metals, including uranium, cobalt, gold and silver, have been reported.[5] However, they are of little more than marginal interest with regard to the problem of metal contamination of food. Of greater interest are those strains of normal pasture plant which develop the ability to grow on toxic soil. These plants are of special significance where regenerated mine tailings or former industrialised sites are used for agricultural purposes. They must also be taken into account when municipal sewage sludge is applied as dressing to the soil.

Metal Tolerance in Fodder and Vegetables

Seeds of most plants alighting on soils containing high concentrations of such heavy metals as lead, zinc and copper will die because of the toxic nature of the soil. But because of the great genetic variability existing in populations of wild plants, there will be occasional individuals with the ability to survive such conditions. The resulting plants may not be quite as vigorous or as well formed as if they were growing on normal soil, but they will survive and develop, particularly as they do not have to compete with other more vigorously growing plants. Extensive studies[7] have been carried out in the UK on strains of the grasses *Agrostis* and *Festuca* which can grow on mine tailings rich in toxic metals. The grasses can accumulate enough metals to be toxic to grazing animals.

In none of the examples so far given of metal accumulator plants is there a very real danger of further transfer of the metals to humans.[8] Animals which die of lead or selenium poisoning after eating contaminated fodder are not likely to be used for human food except in extreme cases.

However, if the plants are used directly as human food then the problem may become very serious. The tragedy which struck the small community of Toyama City in Japan in the 1950s was due to the uptake by rice of cadmium from irrigation water polluted by industry. More than 50 people died and many hundreds suffered from a cadmium-induced disease which resulted, among other symptoms, in decalcification of the skeleton to the point of collapse in the terminal stages. Cadmium contamination still occurs in Japan as well as in other countries. Zinc ores are normally associated with cadmium and when they are smelted cadmium as well as zinc pollution of the environment may occur. In fact, multiple con-tamination from such sources is not unknown. In one study of the effect of wind-borne contamination from a zinc and lead refinery, locally grown corn (*Zea mais*) meal was found to contain 0.9 mg/kg of cadmium and 28 mg/kg of lead. In addition, leaf vegetables had on their outer surfaces over 300 mg/kg of lead, 20 mg/kg of copper, almost 900 mg/kg of zinc and 5 mg/kg of cadmium.[9]

In an earlier study, levels of 8000, 5000 and 50 mg/kg dry matter of zinc, lead and cadmium respectively on the surface of vegetation close to a refinery were reported.[10] No doubt internal accumulation of these metals also occurred. Consumption of vegetables contaminated in this way with levels of metal at least externally well above legally permitted maxima would certainly be a health risk.

METALS IN SEWAGE SLUDGE

Of considerable significance, from the point of view of food contamination by metals, is the practice of applying municipal sewage sludge to agricultural land as a top-dressing. Modern man produces an enormous amount of waste from his domestic and industrial activities. When his needs were simpler and the world was not so crowded, there was little difficulty in disposing of refuse. But now, not only is the volume of waste constantly growing, but with the development of the use of a wide range of harmful substances, including the metals, the waste we produce needs special handling. A great deal of our refuse is deposited in holes in the ground left after extraction of clays and gravels and other minerals; some is dumped at sea and a great deal is converted into sewage sludge and municipal compost in treatment plants. Over a million tonnes (dry weight) of sludge is produced every year in England and Wales alone. Quantities produced in other developed countries are equally impressive.

The idea of using sewage sludge as a dressing on agricultural land and thus converting 'the refuse from your garbage can into plant food' has considerable economic and ecological attraction, not to speak of its political value. Sludge is rich in organic matter, on average about 40 per cent of dried weight. In addition it contains about 2.4 per cent nitrogen and 1.3 per cent phosphates. Modern farming methods are believed to lead to an unstable structure in some types of soil, but this can be rectified by the addition of suitable organic matter. Not only would municipal sludge appear to be ideal for this purpose, but it would also serve as a cheap source of the now very expensive nitrogen and phosphate nutrients required by plants. Unfortunately, sewage sludge contains relatively high and variable levels of trace metals. Indeed, as a result of the use of sewage sludge on agricultural land many incidents of crop failure have occurred due to toxic levels of some of the metals.[11] Most of the metals are due to industrial effluent, but domestic activities can also play an important part. Boron, for instance, appears to come from the use of washing-powders, some of which contain as much as 1 per cent of the metal. The presence of zinc, which is usually among the most abundant of the contaminants, has been attributed in part at least to the weathering of galvanised iron roofs, fences and piping. Table 6 indicates the range of concentrations of a number of metals in typical sewage samples. These are considerably higher than the levels found in normal agricultural land, with more than 300 times as much zinc and 100 times as much boron and copper as there would be available in rural agricultural soil.[12]

When such material is added in even moderate amounts to agricultural land a considerable increase in levels of trace metals results. The added metals, apart from boron, are not easily leached out again. As a result, even after one treatment and without further addition of sludge, the metals can continue to exercise an effect on crops for many years. Studies carried out by Purves and his colleagues in Edinburgh have shown, however, that while such enrichment of soil with trace elements does result in an increased uptake of metals by certain plants, not all metals are taken up to the same extent and there can be considerable differences in levels of uptake between different species of plant. Of all the metals studied, boron appeared to be

TABLE 6

NORMAL RANGE OF METALS IN DRY MATTER OF SEWAGE
SLUDGE[12, 13]

Metal	Content (mg/kg)
Silver	5–150
Boron	15–1 000
Cadmium	60–1 500
Cobalt	2–260
Chromium	40–8 800
Copper	200–8 000
Iron	6 000–62 000
Manganese	150–2 500
Molybdenum	2–30
Nickel	20–5 300
Lead	120–3 000
Scandium	2–15
Titanium	1 000–4 500
Vanadium	20–400
Zinc	700–49 000
Mercury	3–77

the most readily accumulated by most plants, and its uptake was associated with severe toxicity symptoms. However, in the case of beans, most of the boron remained in the leaves and stems and little ended up in the seeds. While copper, too, was accumulated, the increase in content of the leaves was so small that it could not possibly be a danger to human health, even in lettuce and similar leaf vegetables. Zinc accumulation was also limited to well below the toxic level. A study[14] of land which had served for several decades as a sewage farm and was subsequently used for general agricultural purposes showed that while levels of copper, zinc and nickel in the soil were much higher than those of normal agricultural land, plant uptake of copper and nickel was no greater than from normal soils. Zinc, however, was seen to accumulate in lettuce and radish grown in the sludge-treated soil.

Purves, in the report already mentioned, warns that refuse may sometimes contain other metals besides copper and zinc. These include lead, cadmium and mercury, which can give rise to much more serious problems than the less toxic elements just mentioned. Though all pollution problems may not be as acute as those of Minamata in Japan, for example, we must not rule out the possibility that similar episodes can occur elsewhere. A wide range of potentially toxic metals is employed in industrial processes, and under present conditions these elements can find their way into sewage sludge and municipal compost which may be applied to agricultural soil.

THE TRACE METAL CONTRIBUTION OF FERTILISERS AND AGRICULTURAL CHEMICALS TO FOOD

Fertilisers

A report[15] of the National Environment Protection Board of Sweden commented on the fact that, while concern was being expressed by many at the danger of using sewage sludge in agriculture, the far more serious problem of the levels of toxic metals in commercial fertilisers was being largely overlooked. According to the Swedish National Board of Health and Welfare, sludge with more than 15 mg/kg of cadmium should not be used for soil improvement on arable land. The annual application of sludge is, in addition, limited to 1 tonne of dry matter per hectare, which corresponds at the limit to 15 g of the metal per hectare. But cadmium levels in commercial fertilisers often exceed those tolerated for sewage sludge. One type, NPK 16–7–13, was found to contain 18 mg/kg of cadmium; several others had more than 20 mg/kg and one superphosphate preparation had 30 mg/kg. The report showed that, while the overall use of sewage sludge in Sweden for soil improvement in 1973 was about 84 000 tonnes (equivalent to 1260 kg of cadmium if the levels were at the tolerated maximum), total consumption of commercial phosphate fertilisers in Sweden in 1971/2 was 731 480 tonnes, equivalent to approximately 10 000 kg of cadmium. This is more than 10 times the total contribution from sewage sludge. It was also noted that lead levels in fertilisers were high, with approximately 50 tonnes of lead spread on Swedish agricultural land in commercial fertiliser applications in the 1971 season.

Similar findings regarding heavy metal contamination of soils and crops due to use of fertilisers have been reported for Australia.[16] Uptake of cadmium, especially by wheat, from soil polluted in this manner, can be particularly high,[17] but several types of vegetable have been shown to be affected in the same manner.[18] A complicating factor which illustrates the complexity of the problem regarding metal contamination of food crops is reported by Kjellstrom and his colleagues working in the Department of Environmental Hygiene of the Karolinska Institute, Sweden.[17] They point out that it has been shown that an increase in soil acidity results in increased availability of cadmium for uptake by plant roots. In recent years increasing SO_2 pollution, possibly due in part to fumes from combustion of fossil fuels in British power-stations and carried across the North Sea by wind, has brought about a decrease in soil pH in Sweden. This development, the authors note, could result in an increased cadmium uptake in foodstuffs even if soil cadmium levels remained constant.

Agricultural Chemicals

Several metals are used, both in organic and inorganic form, as components of fungicides and other agricultural chemicals. One of the earliest to be so used was copper, as copper sulphate mixed with lime and water and known as Bordeaux mixture. It was sprayed on grapes it is said, originally, to deter stealing before the fruit could be picked for wine-making, but was later found to prevent the growth of mildew, both on grapes and on potatoes. It is now used extensively on a variety of crops subject to fungus attack and, if improperly applied, especially too near harvest time, it can cause contamination of food. Since potatoes are normally peeled before use and the spray is applied to the haulms above ground, Bordeaux mixture is unlikely to be a serious danger to the consumer. The high levels of copper sometimes found in wines, however, can often be traced back to late spraying of the grapes.

A far more serious cause of food contamination is another group of fungicides, which use organic mercurial compounds. Misuse of these has resulted in several cases of mass poisoning. In spite of warnings to the contrary and because of a mistaken belief that washing away a red, water-soluble marker dye would make it suitable for human use, seed grain treated with an organic mercury fungicide was milled and used for bread-making in Iraq in the 1950s. The results were not instantaneous but the cumulative effects of organic mercury intake became apparent within some weeks and many cases of serious and even fatal poisoning resulted.[19] Similar misuse in Pakistan and Guatemala in the 1960s resulted again in poisonings of epidemic proportions. Even when the grain was not eaten directly by humans but was first fed to animals and the flesh used as meat, the alkyl mercury fungicide still exercised its deleterious effect, as was unfortunately discovered by a farmer and his family in New Mexico. These fungicides are still widely used, not only on cereals but on other seed plants such as potatoes, and they are a continuing hazard unless stringent precautions are taken. We are becoming more rather than less dependent on such agricultural chemicals, especially with the development of high-yield cereals, and the need for vigilance to detect contamination of foods by substances of this nature cannot be denied.

Arsenic is used frequently as an insecticide spray in orchards. The most common form of the element is in combination with another toxic metal, lead, in the form of lead arsenate. Normally such compounds are applied well before harvest time, but on occasion the time between spraying and picking is too short and the fruit can carry arsenic and lead contamination to the consumer. This has at time been a problem for cider manufacturers and accounts, probably, for several of the records of high levels of both toxic elements detected in some ciders and perries.

Arsenic is also used in organic form as a contact herbicide and, like any of these compounds if misused, could be a source of food contamination. Another possible source of arsenic in food is the drug arsanilic acid which is sometimes added to animal feeds as a growth promoter. In the UK the Joint Committee for Pesticide Residues keeps a watch on the levels of arsenic in food which may originate from such sources. Its reports do not draw a distinction between organic and inorganic forms of arsenic, but in the light of recent discoveries concerning the differences between the effects of organic and inorganic mercury, lead and other metals, this practice may have to be changed in future years. A recent report on arsenic in food sold in England and Wales, published by the Committee,[20] showed that the highest concentrations are found in seafoods, with levels of 170 mg/kg in prawns and 120 mg/kg in mussels. Of non-marine foods, animal fat had the highest levels with an average of 0.35 mg/kg, with green vegetables slightly lower at 0.3 mg/kg. Most of the latter probably originates from agricultural chemicals and soil, whereas the seafoods may show the results of industrial pollution.

An unexpected side-effect of the use of agricultural sprays which contain copper, lead or other metals is the build-up to phytotoxic levels of the metals in soil surrounding the sprayed plants. This has been observed in orchards which have been converted to other agricultural uses after long periods of fruit-growing.

CONTAMINATION OF FOOD BY METAL-CONTAINING WATER

For many centuries man has been profligate in his use of water. Raw sewage and industrial waste were allowed to pour into rivers and lakes and sea until sometimes the waters became corrupt and pestilence was bred. All the great cities of the world experienced, at one time or another, the foul smell and the disease of polluted water. But eventually realisation grew that water was a valuable resource and, apart from the need to sweeten our rivers and streams from the point of view of quality of the environment, it dawned on responsible authorities that pollution and resulting sickness were preventable. In the middle of the nineteenth century, great sewage and waterworks schemes were begun in several countries. With the implementation of stringent legislation, many once-foul rivers have been returned to a state approaching that before the great population and industrial explosions took place. In some places sea trout and salmon began to appear far upstream, where they had not been seen for a century.

But just as we began to congratulate ourselves of our achievement, the Minamata tragedy occurred in Japan. Mercury which was used as a catalyst in a factory manufacturing vinyl chloride was discharged in waste water into the sea. The metal was picked up and accumulated in organic form by fish and other marine organisms. The fish were caught by local fishermen and consumed by them and their families with serious, even fatal, consequences. The discovery of methyl mercury in fish in American and Swedish lakes showed that mercury contamination of water was not confined to the East. Urgent investigations were undertaken in several countries to assess the degree of contamination and the level of danger, and emergency legislation was passed. It was discovered that large quantities of inorganic mercury were being released into many freshwater areas of North America, and rising levels of mercury were reported in fish caught in the Great Lakes and in other waters of the North-East of the USA.[21] Similar discoveries were made in Sweden, the culprit responsible for the contamination often being wood-pulp and paper-making industries. It became clear that mercury and other toxic metals readily entered the food-chain beginning with contaminated water. Moreover, it is especially significant that many living organisms possess the ability to accumulate these metals, often to levels well above what is legally permitted.

Cadmium contamination of food due to polluted water in Japan has already been mentioned. In Sweden fish caught in a river downstream from a nickel–cadmium battery plant showed high levels of nickel, as well as cadmium, of about 10 times the normal value. A US Geological Survey report of 1970 showed that 4 per cent of all samples collected from rivers and reservoirs all over the country had cadmium concentrations exceeding the maximum allowable level set for drinking-water in the USA ($10 \mu g$/litre).

Mercury and cadmium are by no means the only metals which have been shown to be capable of polluting our food supply after passage through water. Typical of many other works effluents are the levels and types of contaminant released into the River Tees system in the North-West of

England by a steel-making plant. This effluent has between 10 and 15 mg/litre of dissolved chromium, copper, zinc, aluminium and manganese, with 30 mg/litre of iron, and lesser amounts of arsenic, uranium, titanium, vanadium, cobalt, tin, zirconium, antimony and tungsten.[22] The total quantities of some of the metals released from the South Teesside steelworks has been estimated to be (in kg/day) iron 7500, manganese 2300, zinc 850, lead 310 and copper 25, but this is to a large extent particulate matter and much settles out in a lagoon before being released into the surface-water system. Nevertheless, as the first series of figures indicates, a considerable amount of dissolved as well as particulate metals do escape. These figures suggest that not only is there considerable room for a reduction in the pollution load discharged into the water system, but there is also scope for recovery of quantities of valuable metals. In this particular case the British Steel Corporation accepted the need for improved pollution control and undertook extensive measures to achieve this end.

The introduction of legislation in many countries relating to the control of water pollution has generally led to a reduction in the amount of toxic metals released by industry. In Britain, for instance, the Deposit of Poisonous Wastes Act 1972 and the Control of Pollution Act 1974 have had considerable effects on an industry that traditionally has contributed a great deal of metal contamination to water systems. Metal-finishing produces great amounts of toxic and corrosive solutions and metal-rich sludges. Those produced by plating shops especially can be very complex and variable in composition. A typical analysis of sludge and effluent from one plant showed the following composition: nickel 2.5 per cent, zinc 1.5 per cent, chromium and copper 1 per cent, cadmium 0.25 per cent and lead 0.2 per cent.[23] Discharge of such effluents into the sewage system or disposal by other means is, especially since the passing of the Acts, a costly business and, in addition, manufacturers are becoming increasingly aware of the need to respond to public demands for control of environmental pollution. As a result, progress is being made in methods of handling such waste. These include detoxifying and recovery of metals and other valuable waste products. The fact that public opinion is so outraged and government agencies so quick to pounce, when what have been called 'environmental pollution highlights associated with the disposal of industrial waste'[24] occur, illustrates well that the old 'out of sight, out of mind' attitude towards disposal of toxic by-products of industry is no longer acceptable.

Contamination of food by metals in industrial effluent and industry-polluted water is, fortunately, not common. It cannot, of course, be ignored or the possibility of its occurring denied. The Japanese incidents of mercury and cadmium poisoning in the 1950s and 1960s, as well as the evidence provided by the presence of polluted crustaceans and fish in British coastal waters and in American and Swedish lakes, show that responsible authorities must always be alert to the danger that necessarily results from our present industrialisation. It must be remembered that it is a controllable problem where good industrial housekeeping and the operation of well-designed municipal and other water-treatment plants can remove the danger of excessive levels of metals in water used for food production or drinking.

METAL CONTAMINATION OF FOOD DURING PROCESSING

Contamination at the Factory Door

It would be misleading if the treatment of possible sources of metal contamination of primary foodstuffs covered in the previous section were to give the impression that normally a high proportion of food arriving at the processing plant will contain more than a minute amount of toxic or potentially toxic elements. In fact, such occurrences are rare. Current legislation and monitoring practices of public health agencies in most countries see to it that levels of contaminants are normally kept at or below levels which are known as GRAS—'generally recognised as safe'—in the USA. However, most vegetables and many other primary foodstuffs will require some form of cleaning before processing. At the factory gate they may contain particles of soil, dust and other adventitious materials picked up during growth, harvesting and transport. Under certain conditions, to which reference has been made already, dust and soil particles may be rich in metals. If these are not removed they could cause contamination of the processed food. However, it has been found that, less commonly, metal contamination on a large scale might result from the presence of rogue metal—for example, fragments of harvesting equipment or metal containers which have by accident found their way into the food. Such dirt is routinely removed by one or a combination of standard cleaning operations carried out at food-processing plants. Various types of metal detectors are available for the removal of actual metallic fragments. In-line controls can be set up which are capable of detecting the ferrous and non-ferrous metals most frequently encountered as contaminants of this type. These and other types of cleansing equipment and procedures are better treated in texts on food engineering where details will be found.[25]

No matter how free from metallic contamination food is before it reaches the processing plant and the kitchen, there is no guarantee that the end-product which is sold to the public will not contain undesirable amounts of certain metals. Pick-up may occur from the equipment used for processing or from packaging or even as the result of the deliberate use of additives.

Contamination from Plant and Equipment

Food-processing equipment and containers have long been recognised as a source of chemical as well as microbiological contamination of food. It is said that the Romans suffered from chronic poisoning due to leaching of lead from glazed pottery vessels used to store wine.[26] In modern times the use of similar poorly glazed vessels to hold olives during pickling has resulted in lead poisoning in Yugoslavia.[27] Lead has always been a problem in food processing because the metal lends itself readily to the fabrication and repair of cooking and storage utensils. A report from France notes that a traditional method of repairing wine casks by nailing strips of lead with copper nails over cracks is still widely employed, though the practice is known to cause contamination of wine.[29] Similar amateur repair work on cooking and storage equipment, in which solders containing high levels of lead are used to join breaks in metals or attach loose handles, is another cause of lead contamination. These incidents are normally of limited significance and occur more often in small-scale and home production rather than in modern manufacturing plants.

Metals for Food Processing

In modern food-processing plants which have been designed with the need for hygiene and cleanliness in mind, taking the type of materials being processed into account and with awareness of its compatibility with the structural materials to be used, there is little danger of metal pick-up by the products. High-quality stainless steel and plastics approved for contact with foods are used. The actual grade of materials employed will depend on the particular products to be handled. Dairy products, for instance, are particularly sensitive to some forms of contamination and require special consideration. A full treatment of suitable metals and alloys for plant fabrication in the dairy industry is given by Harper and Hall[30] and details for other food-processing industries will be found in Brennan et al.[25] and similar books on food engineering. We will look at the composition of some of the steels and other alloys and plating metals used in the food industry in later chapters when we consider the individual metals in detail.

Whitman[31] has discussed the interactions between structural materials used in food-manufacturing plants, foodstuffs and cleansing agents and the likelihood of their causing food poisoning or quality defects. He points out that arsenic, lead, mercury, cadmium and zinc are all potentially causes of food poisoning. In addition the manufacturer must be alert to the catalytic oxidation reactions and the resulting rancidity of fats, which can be brought about by minute amounts of copper, nickel, iron and chromium. All of these metals may be present in stainless steel, but it is unlikely that with high-quality steel and under good operating conditions they will migrate from processing equipment into food. A study[32] carried out at the Institute of Hygiene in Rome showed that while manganese, chromium, iron and nickel did migrate from stainless steel into food, the level of migration was relative to contact time. Even after 30 days of contact, the amount of the trace metals entering food was very small and far below toxic limits.

Unsuitable Metals
The problem of metal contamination of food during processing occurs frequently as a result of misuse of equipment or the overlooking of the consequences of using unsuitable metals in apparently insignificant ways.[33] Cadmium, for instance, which is often used in engineering to plate small pieces of equipment, is readily soluble in weak acids. Its use should be confined to parts of plant which do not come into contact with food. Unfortunately, cadmium-plated vessels have been used improperly on a number of occasions and poisonings have resulted. Nickel is also used frequently as a plating material and, like cadmium, if equipment containing the metal comes into contact with certain foods, contamination can result. Although nickel is not a toxic metal, it is especially unwelcome in fat- and oil-containing foods, which readily develop oxidative rancidity under the catalytic influence of the metal. Scraped-surface heat exchangers are frequently nickel-coated and, if moved in a food plant to a different duty from that for which they were originally designed, corrosion can occur and contamination may result. Copper, like nickel, catalyses rancidity of fats and is a frequent source of trouble in the food industry. The metal itself, or brass or any alloy which contains it, should never be allowed into contact with fats.

Tinned wire baskets made of brass have been implicated in copper contamination of oils used in deep-fat frying. A copper sampling instrument has been known to cause an incident in which a mere 2 min contact resulted in a fourfold increase in copper content. Brass and bronze taps on equipment have resulted in similar contamination, not only with copper but also with lead, as in the case of a shipment of wine in a tanker which had bronze taps.[34] However, copper contamination and rancidity of food may not always result from the use of copper heating coils and cooking pans in food-oil refineries and confectionery plants where the practice is common. This is probably because the metal is not subject to wear under these conditions of use. In addition a thin film of cooked material coats the surfaces of the metal and provides some protection. Indeed, it is not uncommon for initial problems of metal contamination due to the installation of new equipment to solve themselves with time. This has been observed with zinc-plated galvanised iron trays used for sorting apples in the USA and in Germany. Initially a high level of zinc contamination was observed but the fruit acids quickly formed a chemical barrier by combining with the zinc to give a less soluble salt on the surface of the trays. Iron uptake by food cooked in cast-iron frying-pans and lead extraction from tin plate on tinned copper saucepans have also been shown to be reduced after use. Where copper is used for food-processing equipment it is frequently coated with a layer of tin, which forms a metal barrier between the easily dissolved and chemically active copper and the food. Iron is also treated in the same way, thus preventing corrosion and solution of the underlying, less resistant metal. As we shall see, pure tin is relatively resistant to attack by food acids and, even if dissolved, is non-toxic except at high concentrations. Pure tin is not normally used for plating. An alloy containing a small amount of lead is preferable for practical reasons. As a result, tinned utensils can contribute a not insignificant amount of lead to foods, especially if these are of an acidic nature.

CONTAMINATION OF FOOD IN CATERING OPERATIONS

Tinned copper cooking utensils have been traditionally favoured by chefs. Today, hotel and restaurant kitchens, as well as many domestic kitchens, frequently use tinned copper vessels.

Results from a study of the levels of lead and copper contamination which can be expected from the use of such tinned copper cooking utensils in a restaurant kitchen[35] indicate the extent of the problem. Even though the vessels were within the limit of 0.2 per cent lead in the tinning allowed by UK legislation, it was found that 8–9 mg/litre of the metal were leached from the surface of a series of new saucepans by a 4 per cent acetic acid solution. After a period of use the amount of lead leached was considerably reduced until, in a visibly worn saucepan, it was down to 1 mg/litre of the leaching fluid. Table 7 compares the levels of lead and copper in different types of food before and after cooking in tinned copper and in aluminium

TABLE 7
COPPER AND LEAD CONTENTS OF FOOD BEFORE AND AFTER COOKING (mg/kg)

	Fish		Chicken		Cabbage		Potato		
	Pb	Cu	Pb	Cu	Pb	Cu	Pb	Cu	
Uncooked	0.31	0.82	0.14	2.21	0.15	1.36	0.19	3.10	wet wt
	(1.87)	(6.38)	(1.05)	(17.15)	(0.55)	(7.36)	(0.67)	(10.89)	(dry wt)
Aluminium	0.36	1.37	0.21	2.52	0.18	1.04	0.16	1.87	wet wt
utensils	(1.91)	(7.23)	(1.04)	(6.67)	(0.53)	(5.88)	(0.72)	(8.40)	(dry wt)
Tinned copper	0.42	5.70	0.25	6.36	0.29	2.07	0.22	2.39	wet wt
(old)	(2.00)	(27.01)	(1.62)	(15.35)	(0.66)	(8.01)	(0.84)	(9.07)	(dry wt)
Tinned copper	1.09	2.24	0.94	4.05	0.79	1.93	0.26	1.88	wet wt
(unused)	(4.22)	(11.47)	(3.14)	(13.56)	(3.29)	(5.74)	(1.90)	(7.32)	(dry wt)

saucepans. It is clear that both copper and lead can be leached by the food to an extent dependent on the state of wear of the utensils. After some use the effectiveness of the tin barrier between food and the copper is reduced, allowing solution of the underlying metal. In a similar study,[36] a commercial dried tomato soup was rehydrated and heated in tinned copper saucepans with the following results for lead: previously unused saucepan 3.53 mg/kg; worn saucepan 0.71 mg/kg. The tinning on the new saucepan was found to contain 0.28 per cent lead, while that of the used saucepan had less than 0.1 per cent.

Other metals may also be picked up by food from pots and pans during cooking, sometimes to a toxic level. There is evidence that the use of cast-iron cooking pots by Africans, especially when heated over slow fires for long cooking times, has contributed to an excessive dietary intake and in some cases resulted in iron-overload of the liver.[37] It is probably true that the same has occurred in other simple societies, and has been reported for Papua-New Guinea.[38] Apart from its effect on health, iron uptake from cooking utensils not infrequently causes unwanted colour changes in food. Of course, there may be a compensatory effect resulting in the marked low level of incidence of iron-deficiency anaemia encountered in many parts of Southern Africa.

Just as in industrial processing of food, so in domestic and even commercial catering establishments, the misuse of items of equipment can lead to metal contamination of food. Galvanised iron is sheet iron coated with a layer of zinc to prevent corrosion. The zinc, which usually contains some cadmium, is easily dissolved by dilute acids and should never be used for holding foods which are acidic in nature. Unfortunately, this has not always been recognised by cooks and manufacturers and as a consequence cases of poisoning have occurred. Galvanised buckets have been used to hold soup and meat stock in kitchens and there is a temptation for the home brewer and wine-maker to use readily available galvanised drums to hold his fermentation mixture. The end-product may, as a result, contain high levels of zinc as well as of other metals.[39]

In general, zinc-plated metal should only be used for storage racking, hooks and other items which will not come into immediate contact with wet foodstuffs. An even more restricted use should be made of some other articles made of iron or steel containing certain highly toxic components. Cadmium-plated metal has already been mentioned. In addition beryllium, which is used as a hardener in copper alloys, should be excluded from all possible contact with food. The alloys are highly toxic and even if initially used in a safe position in a food-processing plant, there is always a danger that subsequently the beryllium-containing item might be used for a different duty and food contamination might result.

Uptake of Metals by Food from Glazed Ceramics and Enamelled Utensils
Although limited in scale and seldom of consequence to food manufacturers, metal contamination of foods due to the use of poor-quality glazed pottery vessels and enamel utensils must be included in this study. In Britain alone more than 3000 tonnes of lead are used annually in the glazing of ceramics. The glass-like surface on well-made plates and dishes should normally be unaffected by food and not release its lead and other metal components when in contact with food. Unfortunately, a great deal of glazed pottery, especially the craft and home-made variety, is capable of releasing toxic amounts of lead and other metals into food. There have been several reports of poisonings due to the use of such utensils for storage and preparation of foods and beverages. In an investigation carried out by the British Consumer Association, many earthenware casseroles on sale in high-street shops were found to be poorly glazed and capable of contaminating food. One contributed approximately 200 mg/kg of lead to a stew cooked in it. This was more than 100 times the UK maximum permitted level. A similar investigation in Canada on more than 260 earthenware items showed that 50 per cent of all glazed surfaces tested were unsafe for table use. It may be some consolation for food manufacturers whose plant uses large, glazed ceramic storage tanks and sinks to learn that the more robust and harder stoneware from which these items are normally made is much less liable to release lead into food than are the earthenware

vessels examined in these investigations. From the domestic user's point of view, the best line of action with regard to glazed utensils would be to follow the advice of the UK Ministry of Agriculture, Fisheries and Food inter-departmental Working Group on Heavy Metals (1974) and use only glazed ceramics produced by reputable manufacturers, leaving craft and home-produced pottery as ornaments on the shelf. It is worth noting that because of the wide-scale nature of the problem in the USA a ceramic dinner ware lead surveillance programme has been instituted.

Domestic utensils such as metal casseroles, plates and saucepans which are enamelled may also be the cause of metal contamination of food. Once more those manufactured by reputable makers and which are guaranteed to conform to British or other standards of safety are generally safe. But others of doubtful origin, especially if they be brightly coloured in yellow or red, may contribute both lead and cadmium to food. Enamel is rather like the glaze on pottery: a glass-like layer baked on to the surface, serving both as an anticorrosive lining and as a decoration on the vessel. The colouring matter often contains cadmium and lead salts. Concern has been caused in many countries by the discovery of the presence in stores and super-markets of enamelled cooking utensils which were capable of contaminating food. Standardised tests have been prescribed under food safety regulations in many countries and stringent limits set for the maximum permissible levels of cadmium and lead which can be leached from such utensils. An example of one such regulation is Decree No 408 of 21 November 1962 as amended by subsequent decrees, of the *General Regulation for Foodstuffs* in Finland. It reads in translation:

> 'A receptacle, machine or any other utensil or apparatus used or intended for use when cooking, frying, otherwise preparing or preserving, transporting or consuming foodstuffs in such a way that it may come into contact with these foodstuffs, must be such as not to make the food injurious to health.
> 'Any utensil or apparatus must not be used when the foodstuff could come into contact with lead, zinc or any other similar metal, alloy or substance which may yield poison to such an extent as to be a possible cause of poisoning. The use of such a utensil or apparatus is forbidden if it contains arsenic which will dissolve into the food. No utensil or apparatus may contain lead, zinc, cadmium or antimony to such an extent or in such a form that when the said utensil or apparatus is kept in a 4 per cent acetic acid solution for 24 hours in normal room temperature, more than 0.6 mg of the aforesaid metals

are dissolved in the solution per each full square decimetre of that part of the surface of the utensil or apparatus which comes into contact with the food. Nor must colour be extracted from it in normal use or dissolve in the aforementioned acetic acid solution to such an extent as to change the colour of the solution.'

We shall later note similar regulations from other countries and comment on their contents and the problems which arise from such broad references to quantities of metals which may be 'a possible cause of poisoning'. For the present this will suffice to indicate the concern of public health authorities that food be protected in as many ways as possible from contamination by toxic metals.

Another domestic source of cadmium and lead contamination which has been receiving attention in recent years is the decoration and printing applied to glass tumblers and other containers. For example, a consignment tested in Japan in 1974 released between 0.1 and 390 mg/cm^2 of lead and between 0.1 and 40 mg/cm^2 of cadmium into a test solution of 4 per cent acetic acid.[40] The toxic metals are released almost certainly from the printing and not from the glass itself. Indeed, with the constantly growing practice of selling plastic- and paper-wrapped food, the possibility of contamination of the contents by metal-containing print is a cause of concern at the present time. Polythene bags have been found to contain nearly 25 g of lead per kg. Much of this contamination appeared to be due to the pigments used. However, since metal-containing stabilisers are used in the manufacture of such plastics some of the contamination may come from the plastic itself. Printed paper, too, can contain high levels of lead and even unprinted paper wrappings have been found to contain more than 50 mg/kg of lead. To some extent the lead in wrapping paper may be due to the use of recycled printed paper. Letterpress magazine ink may contain 29 000 mg/kg lead in yellow and 4100 mg/kg in red print. A recent investigation[41] of the lead content of coloured paper packages used for food and confectionery found unacceptably high levels of the metal in the wrappers. A confection wrapper, with mixed yellow, blue, red and brown colouring, had a remarkably high level of 10 125 mg/kg of lead. Not far below this level was a candy-bar wrapper (7125 mg/kg). Lesser quantities were found in food packs. A flour bag had 20 mg/kg and a spaghetti box 50 mg/kg.

It should be pointed out that most foods are actually separated in packs from the printing by an intervening barrier, but this is not true of many sweets, chocolates and candies. In addition, in the course of ordinary domestic use many wrappers, especially of frozen goods, become torn and soggy from moisture and come into contact with the contents. One example of direct contact with food, given by the authors of the report referred to above, was a coloured trading card with 88 mg/kg of lead wrapped in immediate contact with chewing-gum.

Not all types of paper contain such high levels of lead or of other toxic metals. Nor will food grade plastics normally be a source of metal contamination of foods. Indeed, there are regulations in most countries which control the quality of wrapping materials which may be used on foods, and provided these are obeyed there is little danger for the consumer. A study carried out by the Canadian Government Health and Welfare Department[42] makes this clear. X-ray fluorescence scanning of 62 plastic food containers purchased in local supermarkets showed the presence of barium, bismuth, cadmium, copper, iron, mercury, nickel, lead, selenium, strontium and zinc. However, no lead and only 0.02 μg/cm^2 of mercury and traces of cadmium from 0.002 to 0.02 μg/cm^2 were leached when the containers were tested for the solubility of these three toxic metals. Nevertheless, further tests did show that exposure to ultraviolet light and surface abrasion rendered both cadmium and mercury more soluble. They also indicated that, if ingested, small fragments of some of the plastic could release upwards of 0.266 μg/cm^2 of cadmium.

Apart from contamination of food due to the presence of metals in printing inks and dyes used on food-packaging materials, the presence of dyes and other colouring substances directly in foods themselves has been known to cause problems. The use of food colours is controlled in most countries by food safety regulations and the need for the establishment of international specifications for the identity and purity of food colours has been stressed by the Joint FAO/WHO Expert Committee on Food Additives. The need for such control was underlined by a recent report[43] from India. Of 36 food colours examined, 18 were approved by the Indian Government and the other 18 non-approved but nevertheless widely used in commercially available foodstuffs. All 36 were found to contain arsenic, cadmium, chromium, cobalt, copper, lead, manganese, nickel and zinc. Levels of arsenic, chromium, copper and lead in the permitted colours were found to conform to legal limits, but many of the non-permitted colours were in excess of these limits. In some, the level of arsenic, for example, was more than twice the allowable limit of 3 mg/kg.

CONTAMINATION OF FOOD BY METAL CONTAINERS

In a later section the question of the preservation of food by canning will be discussed in detail, with particular reference to the use of tin and aluminium, but a brief look at some possible types of contamination of foods in metal containers will help to round out the answer to the question, how do metals get into food? While it is true that every process to which food is subjected and each vessel or container in which it rests makes some contribution, big or small, to the load of chemical contaminants which it ultimately carries into our bodies, packaging is in fact only one source among all the others.[45] Nevertheless, for historic reasons at least, contamination of food by metal from cans looms large in public awareness. There is plenty of evidence that when tin plate was first used to make containers for food about 150 years ago, many cases of food poisoning, apparently due to ingestion of excessive amounts of metal, occurred. A congress of physicians held in Heidelberg in Germany even went so far as to recommend that 'tin plate should be forbidden for the making of vessels in which articles of food are to be preserved'.[46]

The quality of tin plate has been greatly improved since those days. No longer is unprotected tin plate used exclusively for cans. Organic lacquer coatings—mainly thermosetting resins which are polymerised on the surface of the sheet tin plate by a baking process—are used as a barrier where there is a danger of the contents corroding the metal of the can. However, in the months or even years during which cans may be stored, a very slow corrosion process frequently goes on inside the can, resulting in an ever-increasing level of tin, iron and sometimes of other metals in the food. Various factors such as the amount and type of organic acids present, the levels of nitrate, the amount of oxidising or reducing agents, storage temperature and the presence or absence of lacquering, determine the amount of corrosion. In addition, in many cans solder is used on the seams and this contains a high proportion of lead. In modern cans solder is normally exposed to food in two small areas only, in the top and bottom. Sometimes, however, solder splashes are found elsewhere inside the can. In unlacquered cans it is relatively easy to remove these solder splashes before the can is filled, but it is not so easy to do so with other types of can. Lead-containing particles may remain stuck to the lacquer and come into contact with the can contents. Solder was used much more extensively in the construction of earlier types of can and there was far greater contact between solder and food.

It is of interest, as well as an illustration of the electrochemical principles of the cathodic behaviour of lead relative to tin (i.e. when both metals are in electrical contact, the tin will dissolve in preference to lead and lead already in solution will be replaced), that analysis of a tin of roast veal which had been taken on an Arctic expedition in 1837 but was actually opened in

London 100 years later, was found to contain 71 mg/kg of iron, 783 mg/kg of tin and only 3 mg/kg of lead. A similar tin of carrots in gravy contained 308 mg/kg of iron, 2440 mg/kg of tin and no lead.[47] This electrochemical behaviour is, of course, an important factor in the acceptability of canned food. However, in spite of improved techniques and materials, problems are still experienced with regard to metal pick-up in canned foods. A study of canned Israeli fruit juice[48] showed over 700 mg/litre of tin as well as more than 2 mg/litre of lead in some samples. The authors of the report commented on the fact that some cans on sale in retail shops were more than three years old and that these were richest in metal contaminants.

A report from New Zealand[49] compared the amount of lead contamination in several types of canned food, including baby food, available in retail shops. Differences were found between lacquered and unlacquered cans. The authors concluded that generally the incidence of lead contamination is significant and is mainly associated with lacquered cans. This view receives some support from results published by the Laboratory of the Government Chemist in the UK.[50] Considerable inter-can variation for lead, tin and iron was found. While blackcurrants in a lacquered can contained 10, 160 and 2600 mg/kg of lead, tin and iron respectively, rhubarb also in a lacquered can had 0.3, 30 and 24 mg/kg of the same metals.

In contrast, gooseberries in plain cans had 0.3, 260 and 4.3 mg/kg and similarly canned pineapple 0.35, 105 and 3.0 mg/kg of the metals. Similarly green beans (lacquered can) had 0.7, 10 and 4.8 mg/kg and broad beans (plain cans) had 0.02, 10 and 18 mg/kg of lead, tin and iron. The effect of storage temperature with time on the tin content of canned green beans was studied by Catala and Duran.[51] They showed that while tin plate behaved satisfactorily at approximately 20°C, at 37°C electrolytic tin plate could be completely detinned over a two-year period, with the tin content rising to over 100 mg/litre in the liquid and 1000 mg/kg in the beans. The effect of leaving an open can exposed to air at room temperature can be to raise the level of tin entering the food from the tin plate. Studies carried out by Kimura et al.[52] have shown this for baby foods. The amount of nitrate present in the food has also been shown to increase tin uptake.[53] It has been suggested that the presence of nitrate in addition to tin increases the toxicity of the metal.[54] An outbreak of sickness in the Saar district of Germany was traced to the consumption of canned peaches produced by a small Italian company. The fruit was found to contain about 400 mg/kg of tin, compared to the general level of 44–87 mg/kg found in canned peaches produced elsewhere. In addition, the suspect cans contained a high level of nitrate. This was traced to the use of well water which had upwards of 300 mg/litre of NO_3. It was concluded by the investigators that the level of tin in the peaches was not of itself likely to be toxic, but in combination with nitrate was responsible for the illness.

Apart from tin, iron and lead, which are the main constructional metals in cans, several other metals may occur on occasion in foods preserved in metal containers. Zinc and copper are frequently encountered and permissible levels of these metals in canned food are being considered by some countries. Austria has proposed 50 mg/kg for copper and 100 mg/kg for zinc—levels which are in excess of most results reported for all but the most seriously contaminated foods. A Swiss study for instance found 0.4–7.3 mg/kg of zinc in fruit and vegetables stored in cans, which was not very different from levels of the metal in the same foods packed in glass jars.[55] This finding suggests that the zinc contamination occurred before packaging.

As has been mentioned already, the tin used in tin-plating is not always 100 per cent pure. In the UK tin plate may contain 0.2 per cent lead. In other countries different standards apply. For example, Polish legislation specifies 99.75 per cent purity of tin. The non-tin fraction may contain upwards of 0.05 per cent arsenic. As a result of using such tin in canning, between 0.244 and 0.380 mg/kg of arsenic could penetrate into the food, depending on can size.[56]

Aluminium Containers

Aluminium containers for food have been available for more than half a century. Initially they caused many problems because of technological difficulties and were not widely used. However, in recent years, with advances in metallurgy and container-making processes and a better understanding of the chemistry and patterns of corrosion of the metal, the use of aluminium for food packaging has increased dramatically. The metal is employed in the food industry in a variety of ways, as metal cans, consumer foil, flexible packaging, caps and closures. It is fast becoming one of the most common materials used for food packaging.

One of the main attractions of aluminium as a packaging material is that it is recognised by responsible authorities as safe at the levels at which it is normally ingested with food. However, as with most other packaging metals, the pure metal is not generally used commercially. To provide strength, improve formability and increase corrosion resistance, various alloying elements are added. These include iron, copper, zinc, manganese and chromium. These metals, as well as aluminium itself, may migrate into the contents of a container if corrosion takes place. Uncoated aluminium cans are readily attacked by many fruits and vegetables. The metal may dissolve in the liquids and hydrogen gas evolve, leading to can swell. In addition, discoloration and flavour changes may occur. Consequently aluminium cans are often coated to prevent direct contact with the contents.

Aluminium cans are increasingly being used for both alcoholic and non-alcoholic beverages. It has been shown that levels of between 5 and 10 mg/litre of dissolved aluminium in soft drinks will not affect the flavour and that beers especially in aluminium cans are superior in flavour, colour and clarity to beer packed in tin-plate cans.[57] In fact, the presence of aluminium in conjunction with tin plate or tin-free steel can have a protective effect on the quality of canned beer and other beverages. The aluminium has been shown to act as a 'sacrificial anode' and stops the reaction between the can contents and the iron or tin. Aluminium ends on tin-plate or steel can bodies provide electrochemical protection to the contents in this manner.

Nevertheless, it is generally found that beers as well as soft drinks have a longer shelf-life if aluminium cans are lined with vinyl epoxy or some other resin. Strong alcoholic drinks such as whisky and brandy as well as wines can cause undesirable reactions with aluminium if it is uncoated. For example, pitting corrosion may take place and aluminium pick-up can cause discoloration and the formation of a flocculent precipitate of aluminium hydroxide.

Aluminium foil is extensively used for food wrapping. If properly handled and used in the right conditions, it provides an excellent and highly acceptable form of packaging. However, in the case of frozen foods for example, bad handling during distribution, resulting in thawing, can cause both deterioration of the food and corrosion of the container. In addition, foil-wrapped foods will remain satisfactory as long as they are in contact only with non-metals, but if they touch another metallic object such as a steel or tin-plate vessel, an electrochemical reaction takes place with, once more, the aluminium acting as sacrificial anode. A rapid pitting corrosion of the foil will take place and the quality of the food can be affected adversely.

CONTAMINATION OF FOOD BY RADIOACTIVE METALS

Becquerel discovered in 1896 that the element uranium emitted rays which could affect the emulsion on photographic plate. In 1908 Madame Curie and her husband Pierre showed that the atoms of certain elements such as uranium and radium undergo spontaneous disintegration to form atoms of other elements and that they emit penetrating rays in the process. These spontaneous decompositions of atomic nuclei are what we call radioactivity. Several radioactive elements occur in nature and man has been exposed to their penetrating rays to a greater or lesser extent ever since he first evolved on earth. But during the past few decades, a new form of radioactivity, artificially produced, has been introduced into the environment. Since the 1940s, following the invention of the atomic bomb and the harnessing of nuclear energy, artificial radioactivity has become a widespread hazard contributing sometimes to the problem of food contamination.

The penetrating rays which are emitted by radioactive elements are commonly referred to as ionising radiation because of the effect they produce. They are high-velocity particles emitted from the nuclei of the atoms. Two principal kinds of particles are given off: α-particles, which consist of two neutrons and two protons and which are identical with helium nuclei; and β-particles, which are high-velocity electrons. In addition, these particles are frequently accompanied by γ-rays. These are electromagnetic rays, like X-rays in character, and are more penetrating than either the α- or β-particles. When an atom suffers the loss of an α-particle, its atomic mass will be decreased by four units and its atomic number by two. The loss of the two protons from the nucleus is accompanied by the loss of two electrons, so that electrical balance is maintained. Thus, for example, the element radium disintegrates with the loss of an α-particle to give a new element radon, as follows:

$$^{226}_{88}\text{Ra} \rightarrow {}^{222}_{86}\text{Rn} + {}^{4}_{2}\text{He}$$

Radon in its turn spontaneously disintegrates with the loss of another α-particle to give polonium:

$$^{222}_{86}\text{Rn} \rightarrow {}^{238}_{84}\text{Po} + {}^{4}_{2}\text{He}$$

When a β-particle is emitted, the result is as if a neutron in the nucleus decomposes to give a proton and an electron; the proton remains in the nucleus but the electron is emitted. As a result there is an increase of one in the net positive charge on the nucleus so that the atomic number increases by one unit, but there is no appreciable change in the mass of the nucleus. An example of this change is the transformation of an isotope of thorium into one of protactinium:

$$^{232}_{90}\text{Th} \rightarrow {}^{232}_{91}\text{Pa} + {}^{0}_{-1}\text{e}$$

Emission of either α- or β-particles may be accompanied by γ-radiation. The emission of γ-rays has no effect either on the atomic number or on the atomic weight of an atom, since these rays possess neither mass nor charge. All elements with atomic numbers larger than that of bismuth have one or more isotopes which are naturally radioactive. A few elements of low atomic number, such as potassium and rubidium, also have naturally occurring isotopes which are radioactive. Artificially produced radioisotopes of many other atoms are also produced.

The rates at which disintegration of radioactive elements takes place vary between elements. The number of atoms that disintegrate in unit time is a constant fraction of the total number of atoms present. The time required for one half of the atoms in a sample to disintegrate is known as the half-life. While in the case of radium the half-life is 1590 years, it is only 3.82 days for radon and 3 min for polonium.

When the Curies began work on radioactivity, no one suspected that such work was hazardous for health. Unfortunately, it soon became evident that radiation of this type could result in sickness and death. We know far more about the consequence for health of nuclear radiation today, especially after the tragic incidents of over-exposure that have occurred since the development of atomic weapons. It is a cause of concern today that through radioactive contamination of food very many people may be exposed to such radiation.

There are, in fact, many sources of ionising radiation in the world. Indeed, as it has been well put[58], 'We live all our lives in what has been called a "sea" of such radiation. Known as the natural background, this radiation comes from cosmic rays, from radioactive substances in the earth's crust, and from other such substances (such as potassium 40) that circulate through the living world. It amounts to an average of from 0.08 to 0.15 rad per person per year (a rad is equal to the energy absorption of 100 ergs per gram of irradiated material). That man has evolved in the presence of this inescapable background does not mean that it should be regarded as "safe", however, nor does it mean that we should take lightly any man-made additions to the background that happen to be smaller than or comparable to it.'

One of the most important man-made additions to our level of exposure has resulted from the application of nuclear fission to warfare and energy generation in the past four decades. It is from this application that most of our radioactive contamination of food results.

Radioisotopes in Food
Three different types of natural radioactive elements can occur in food.

Some, such as uranium, thorium and potassium, have always been in the lithosphere. All three have extremely long half-lives of millions of years. A second group is composed of the daughter elements resulting from disintegration of the long-life elements. Radium 226, produced by uranium 238, is one of these. Radium in its turn has a daughter of its own—radon, which is also unstable. From radon are generated two other radioactive elements—lead 210 and polonium 210.

A third type of radioactive isotope is constantly formed in the atmosphere by the action of cosmic rays. One important product of the action of these rays is carbon 14, made by the transmutation of nitrogen. Carbon 14 and the other natural radio-elements present in soil, water and the air, like their non-radioactive counterparts, can enter food-chains and eventually be taken into the human body. There they account for about one-quarter of the total background dose of radiation to which the body is exposed. Potassium 40 is responsible for the major portion of this internal dose of radiation, with considerably less coming from carbon 14.

The artificial radioisotopes which get into our food have increased greatly in quantity since the Hiroshima bomb of 1946. The increase was especially rapid during the 1960s when the nuclear powers carried out many atmospheric tests. In addition, smaller amounts of artificial radioactivity are released into the environment from nuclear power-stations and associated processing plants. A smaller contribution is made by the wastes from medical and research laboratories.

When an atomic bomb explodes or a nuclear reactor operates, a complex mixture of many different radioactive elements is produced as a result of fission of the fuel materials. These include heavy elements such as uranium 235 and plutonium 239. In addition, other radioactive elements are produced as a result of fission, such as strontium 90 (half-life 28 years), cobalt 60 (half-life 5.27 years), ruthenium 106 (half-life 1 year), caesium 137 (half-life 30 years) and the non-metal iodine 131 (half-life 8 days). There are also many different activation products, including zinc 65 (half-life 245 days) and the non-metal carbon 14 (half-life 5760 days). These activation products are not produced directly in the fission process but are formed at the same time by the action of atomic radiation, mainly neutrons, on elements naturally present where the explosion occurs or forming part of the structure of the weapon or reactor. The radioisotopes used in medicine, industry and research are also produced by 'neutron bombardment' of selected target atoms in specially designed reactors.

The years between 1950 and 1962 saw a great increase in levels of radioactive contamination of the environment following intensive testing of atomic weapons. The rise was especially rapid in 1961 and 1962 and public opinion, supported by the informed concern of many scientists, built up in many countries to bring pressure on governments to call a halt to testing. The result was the 1962 moratorium on surface and atmospheric testing of atomic weapons. Since then there has been a marked reduction in the levels of environmental radiation. Nevertheless, escapes of fission products from underground explosions, as well as atmospheric testing by countries that have not agreed to the moratorium and emissions from nuclear reactors, have all contributed to the pool of radioactive elements remaining from the early 1960s. As a result we now have a permanent source of radioactive contaminants of foods which is a far more serious threat to human health than the naturally occurring potassium 40 or carbon 14 ever were.

Contamination of Agricultural Produce

Two of the most important fission products in the radioactive pool are the relatively long-lived strontium 90 and caesium 137. Other important, though shorter-lived, radioactive elements in the pool are ruthenium 106 and iodine 131. In the atomic tests of 1961 and 1962, most of the fission products were released into the upper atmosphere as an aerosol. From there they were transferred slowly to the lower atmosphere, from which rain washed them down to the surface of the earth. When radioactive rain falls, it contaminates both the external surfaces of plants and also the soil, from which it may eventually enter plants through their root systems. External contamination is always encountered at times of high fall-out following atomic weapon testing or a reactor accident and usually involves mainly the short-lived iodine 131. This was the case following an accident at the UK atomic plant at Windscale in 1957: contamination of herbage was sufficient to cause an unacceptably high level of contamination in milk from cows grazing within a 20 mile radius of the plant. However, the contamination was short-lived. A similar pattern of contamination was caused by a controlled nuclear explosion in the Northern hemisphere in September 1976. A French report[59] noted that following the blast contamination of a number of foods had occurred. In particular, contamination in milk and bovine thyroid glands was intense but brief. Monitoring was continued for approximately 5 weeks, though it was considered that the dangers to human health from this specific incident were minimal.

The second route of transfer of contamination, through the soil and plant root system, is more important for the long-lived radioactive elements. Uptake by plants of the various elements is, as we have already seen for the stable elements, related to the nature of the soil, its organic and other components, as well as to the chemistry of the particular element in question. Caesium, for instance, is absorbed very tenaciously by soil organic matter and clay and under some conditions is virtually unavailable to the plant. Strontium, on the other hand, is less tightly bound by soil particles and is relatively easily absorbed by plant roots. In this way it can reach the foliage and be consumed by grazing animals and by humans. Cereal grains become contaminated in both ways. In fact, cereals are probably the most important plant source of human contamination by radioactive elements. The level of activity of caesium 137 in barley and wheat measured in Denmark in 1963 was approximately 100 pC shortly after the beginning of the test moratorium. Three years later levels had fallen tenfold, as weapon testing declined.[60] Other reports have confirmed that levels of cereal contamination have levelled out in recent years, with small peaks due to sporadic testing by non-moratorium powers. The distribution of the radioactivity within the cereal plant is not uniform, for a good deal of the contamination is on the outer tissues. In fact, milling of flour to a 70 per cent extraction rate will remove approximately 60 per cent of the caesium 137 and some 90 per cent of strontium 90 present in the whole grain. Wholemeal flour is second only to animal produce as a source of these two fission products in the human diet. Leafy vegetables may also be an important source of radioactive contamination at times of heavy fall-out, but root vegetables and fruit are usually far less important.[61]

The major dietary sources of the long-lived radioactive elements are foods of animal origin, especially milk and other dairy products. The level of caesium in milk, for example, has been used to provide a measure of total dietary contamination and in many countries monitoring of milk for radioactivity is regularly carried out by the authorities. The high levels in animal products are mainly a result of contamination of grass and grain. A pattern of contamination similar to what we have seen in cereals has been noted in dairy products. After a peak in the early sixties, there has been a significant decline in recent years, with a slight upturn corresponding to the renewal of weapons testing on occasion.

Apart from iodine 131 and carbon 14 which, though not metals or metalloids, must be mentioned in passing because of their importance as radioactive contaminants in food, some other radioactive elements have also been released by nuclear reactors and have entered food-chains in recent years. The man-made element plutonium 239 is an activation product of uranium 238 produced in large quantities in nuclear reactors. It is itself capable of nuclear fission and has been used to make weapons. It is also used as a nuclear fuel in fast reactors. It has a long half-life of 24 400 years. Plutonium 239 has been distributed in the environment by a number of occurrences since it was first used in a nuclear weapon in Japan in 1946. In 1966 a US Air Force bomber was in a collision over southern Spain and three plutonium 239-containing nuclear bombs were dropped near the village of Palomares. Fortunately, a nuclear explosion did not take place, but the conventional explosive used as a trigger device did disperse the plutonium 239 over several fields used for growing tomatoes. A crash of another nuclear bomber in Greenland resulted in plutonium 239 contamination of sea-water. A second isotope of plutonium, plutonium 238, is also used in nuclear devices, especially as a power source in artificial satellites. One such device re-entered the atmosphere above the Indian Ocean in 1964 and in the resulting burn-out scattered plutonium 238 widely throughout the southern stratosphere.

Fortunately, though plutonium, which emits a-rays, is extremely hazardous if directly ingested into the body, its compounds are only taken up by plants with difficulty. Its salts are even less soluble than those of caesium and strontium and therefore it is unlikely to enter food by this route. The greatest cause of concern is that surface contamination of the food plant may occur and thus enter the body by ingestion or inhalation.

Nuclear reactors and weapons can also cause radioactive contamination of foodstuffs in another way. This is through pollution of rivers and oceans, especially by the discharge of waste water containing even low levels of activity. Normally contamination of the ocean, such as occurred off Greenland with plutonium 239, has little effect on marine foods because of the rapid dilution that occurs. But in localised areas, the situation may be quite different. Discharge from the Windscale plant in the UK is carried by pipeline into the Irish Sea off the coast of Cumbria. The waste consists of low-level fission products. Monitoring of marine animal and vegetable life is regularly performed by the Fisheries Radiobiological Laboratories.[62] So far there has been no cause for serious concern with regard to build-up of contamination in fish or crustaceans in the surrounding sea. However, as is the case with mercury and several other metals, there is evidence that some forms of marine life are capable of concentrating caesium 137 in their tissues. Plaice, one of the important species of fish caught commercially in the north Irish Sea, has been shown to concentrate the caesium 137 by a factor of approximately 20 over the water concentration level. An edible seaweed, *Porphyra umbilicalis*, which is consumed in Wales as laverbread, also accumulates caesium 137. The alga has also been shown to be capable of concentrating another radioisotope, the short-lived ruthenium 106, by a factor of as much as 1000. This accumulation takes place normally close to the outfall. Zinc 65 has been found to be accumulated in the flesh of oysters near the discharge pipes of another nuclear plant in the UK. In addition, caesium 137 has been reported in shrimps collected near a number of nuclear plants which are permitted to release waste water into the sea. In another, discharging into fresh water, contamination of trout with caesium 137 resulted. Reports from the USA, Sweden and some other countries indicate that contamination of food by discharge of waste from nuclear reactors or by fall-out has occurred from time to time.

A report from the USA[63] gives evidence which indicates that the level of radioactive contamination of food generally is low. The report, issued from the Federal Bureau of Radiological Health, summarised results of analyses of strontium 90, caesium 137, iodine 131, ruthenium 106 and potassium 40 in samples of the total diet and selected imported foodstuffs for the years 1973 and 1974. Low levels of activity which were well within the radionuclide intake guidelines of the Federal Radiation Council were found. Moreover, levels of potassium 40, for instance, in meat and fish, agreed with the normal levels for 'reference man' reported by the International Commission of Radiation Protection.

3

Quality Control

ADULTERATION OF FOOD

'No person shall sell to the prejudice of the purchaser any article of food or any thing which is not of the nature, substance or quality demanded by such purchaser.' With these words the Sale of Food and Drugs Act 1875, the first effective Act of Parliament relating to food quality passed in the UK, laid the foundation for all modern food laws, both in Britain and in countries that have inherited her system of legislation. It is to meet the right of the purchaser for food of the nature, substance and quality to which he is legally entitled, that the purveyor of food must see to it that his products conform to the laws and regulations relating, among other factors, to metallic contaminants. In other countries where the legal system owes little, if anything to Britain, similar laws and controls govern the manufacturer's practices and protect the rights of the purchaser. In the future we can expect to see universal food laws binding on all member nations, passed by such international bodies as the United Nations and the European Economic Community.

The purchaser has not always been so well protected. There is ample historical evidence to show that legislation to control levels of contamination of food has been sorely needed in the past. Before the days of refined analytical procedures, the problem of main concern with regard to metals in food for example, was their use and that of their salts as adulterants. These were normally of two types: either inert material incorporated into a product in order to increase its bulk and hence the trader's profit, or fraudulent additions of substances to improve the appearance of the product.[1] Frequently chalk (natural $CaCO_3$) and alum (naturally occurring crystalline potassium aluminium sulphate,

$K_2SO_4.Al_2(SO_4)_3.24H_2O$) were incorporated into bread; copper sulphate crystals were put into beer and 'Prussian blue' (potassium ferric ferrocyanide, $KFe(CN)_6$) into tea to improve colour. Other metallic adulterants used even well into the latter half of the nineteenth century were soapstone, talc and French chalk (which are all forms of magnesium silicate), calcium sulphate, lead chromate and iron oxides. Early volumes of the *Analyst*, which was begun as a 'monthly journal devoted to the advancement of analytical chemistry' by the Society of Public Analysts in 1875 and still today fulfils this useful function, carried many examples of these fraudulent, though sometimes merely ignorant practices. It is of particular interest as indicating that such practices occurred not just in Britain that the *Analyst* often included reports abstracted from Continental journals. In 1891, for instance, a paper by J. Mayrhofer of the Bavarian Chemical Society on the addition of copper salts to preserve the green colour in cooked vegetables was abstracted from the German *Chemische Zeitung*.[2] Several other papers and letters on this topic appeared in the *Analyst*[3] around the same time, including one which mentioned that there had been several prosecutions in recent years for selling peas to which copper salts had been added. Cheese, apparently, was also adulterated at times with metallic salts. A report[4] of a meeting of the Society of Public Analysts in 1897 noted that metallic lead had been found in a sample of Canadian cheese and that in Britain a substance known as 'cheese spice' was often used to prevent heaving and cracking of the product. This 'spice' was found to consist of 48 per cent crystalline zinc sulphate and it imparted a high level of zinc to the cheese. In addition, copper sulphate was sometimes added. The report continued, 'It is well known that the green mould in certain kinds of cheese has been imitated by the insertion of copper or brass skewers.' Some even more serious cases of cheese adulteration were also noted: 'Instances have occurred in which preparations of arsenic have been added to cheese as a preservative, and in 1841 several persons were poisoned by these means. A similar case occurred in 1854 when a Parisian family suffered, but not fatally.'

It is of interest that the analysts of some hundred years ago were aware of a serious problem which is still with us today. They noted that the use of coloured wrapping materials, such as yellow cheese cloth and papers on 'chocolate, bonbons, etc.' contained lead chromate.[5] Another adulteration problem of current interest at the time, according to the pages of the *Analyst*, was the use of barium salts in cayenne pepper which, an abstract of a paper taken from a German journal states, 'in order to give the substance a brilliant fiery colour, is adulterated with a barium-ponceau lake'.[6]

Though not a food, but rather a medicinal substance, it is worth mentioning that adulteration of opium was also noted, this time in an abstract from the *American Journal of Pharmacy*. In addition to wheat starch, used to increase the apparent bulk of the opium powder, strontium sulphate had also been detected. The *Analyst* notes that 'the author considers its presence particularly objectionable since it causes the percentage of morphine to appear considerably higher than is really the case. It was first noticed about a year ago, and the practice is still continued, some of the samples assayed during the past few months having yielded unusually high results when examined by the United States Pharmacopoeia process.'[7]

These few examples can serve to give some idea of the problems caused by direct adulteration of foods and related materials with metallic salts a hundred years ago. There were, in addition, numerous references to the sort of contamination frequently met even today due to the use of unsuitable material for processing and packaging food: lead from tin plate, solder and pewter, tin from unlacquered cans, lead from rubber rings used to close cans and jars.[8] These instances all occurred well after regulations and laws to prevent adulteration and contamination of foods had begun to be promulgated and enforced. The far more serious situation which existed before those laws were first introduced may be glimpsed in the classic work by F. Accum published in London in the second decade of the nineteenth century. His *Treatise on Adulterations of Food and Culinary Poisons*[9] is well worth perusing by anyone who thinks that today's consumer does not get a fair deal from the food manufacturer!

THE ORIGINS OF BRITISH FOOD LAWS

Britain, like other countries before her, had laws for the protection of the purchaser of food many hundreds of years ago. It was a duty of justices when holding assizes in different parts of the kingdom to inquire into complaints of citizens concerning the quality of bread and beer and other foodstuffs sold by merchants and to punish those who sold adulterated products and underweight quantities. An Act to regulate the sale of bread was made under Henry the Third in 1266. Between then and the nineteenth century many other acts and statutes were passed concerning adulteration and fraud in specific foods. With the massive increase in population following the Industrial Revolution and the shift from rural to urban living, resulting in a corresponding increase in demand for the services of middlemen in the supply of food, the earlier simple laws no longer proved effective in preventing adulteration and fraud of many kinds. The new power of the printed word and pressure generated by what might now be described as the scientific lobby helped to bring about much needed reform in the food laws. Of considerable importance in making such reform effective was development that took place in analytical chemistry in the late eighteenth and early nineteenth centuries.

A select committee on adulteration of food was set up by Parliament in 1855 and its report led to the passing of the first Adulteration of Food and Drink Act in 1860. It was an ineffective piece of legislation, mainly because it left the responsibility for food quality a local and purely optional one. It was improved by a second Act of the same name passed in 1872, which made the appointment of public analysts mandatory. But the law was still defective and it was only in 1875, after another select committee had reported, that the Sale of Food and Drugs Act which is the basis for modern British law was passed. This Act had teeth and offenders could now be fined for the adulteration of food. Subsequently, improvements in the quality of food sold to the public and a decrease in the amount of adulteration detected were considerable. There followed a rapid and sustained decline in the number of prosecutions brought under the Act. Various regulations dealing with the composition and labelling of specific foods were introduced after the First World War. The 1875 Act was repealed and replaced by a number of new Acts during the following half-century, until in 1955 the present Food and Drugs Act was passed. All foods sold in Britain today are subject to the general regulations of this Act. Under it, government ministers have powers to make regulations covering the composition of food, including additives and contaminants. As we shall see, in this way the Minister of Agriculture, Fisheries and Food has been enabled to issue regulations concerning arsenic, lead and copper in various foods.

Two important advisory committees assist the ministers in preparing regulations under the Act of 1955. These are the Food Standards Committee (concerned with the composition, labelling and advertisement of food) and the Food Additives and Contaminants Committee (which advises on the need for and safety in use of additives and the levels of contaminants that should be permitted in food). Regulations are only made after long and wide-ranging discussion with the interested bodies, including scientists and industrialists as well as the consuming public. The laws when promulgated are enforced by food and drugs authorities of the different local authorities. They must appoint public analysts and environmental health officers and by warning, advice and assistance, as well as by prosecution in the courts when necessary, see that the provisions of the law are observed. As R. F. Giles, under-Secretary in the UK Ministry of Agriculture, Fisheries and Food, observed at the symposium celebrating the centenary of the 1875 Sale of Food and Drugs Act, this type of local authority responsibility for the operation of the law is not common to all countries, but, he claimed, 'the United Kingdom system of the law being written by central government and enforced by local government is beneficial to all'.[10]

THE ORIGINS OF FOOD LAWS IN THE USA

Though food laws in the USA may operate in a somewhat different manner from those of the UK, they do have a common origin to some extent with British legislation. As Alexander M. Schmidt, Commissioner of Food and Drugs, US Department of Health, Education and Welfare, remarked at the Symposium referred to above, there is a distinct familiar resemblance between the food laws of the two nations.[11]

In the colonies of New England the law was that of England with 'assizes of bread' for example, corresponding to the practice of the mother country. But, with the developing spirit of independence and still more so after 4 July 1776, the local laws dealing with specific American problems were enacted Since many of these came in response to the needs of producers and consumers in the 13 separate colonies and, later, States, a diversity of laws emerged. One such law, for instance, was that passed in 1723 in the Massachusetts Bay colony which prohibited the distillation of rum in stills in the construction of which lead was used. This prohibition arose directly from the realisation by the colonists that the painful affliction of rum drinkers which they knew as 'dry gripes', similar to the 'Devonshire colic' of cider drinkers in England, was related to the presence of lead in the beverage. It is interesting that some 200 years later 'dry gripes' and other symptoms of lead poisoning appeared among American drinkers who consumed illicit 'bath-tub gin' made in crude stills.

A post-independence law passed in Massachusetts in 1784, the first general food law in the USA, showed that New Englanders took the question of food quality seriously, though no doubt this law was no more effective than the earlier one in stopping abuses. This new law was a general one and provided for the prosecution of 'evilly disposed persons' who 'from

nd filthy lucre, have been induced to sell diseased,
us or unwholesome provisions to the great nuisance of
peace'.

tates continued to pass their own food laws during
ies and attempts to pass national laws in the federal
ot widely supported. 'Many Americans, including mem-
s, thought that food and drug protection, if needed at all,
or the States, and most States already had laws dealing with
Nevertheless vigorous attempts were made to have compre-
aws enacted by Congress, especially after the passing of the
360 and 1875. Between 1880 and 1900 more than 100 bills were
n Congress to regulate food and drugs. A number of minor
iccessfully passed, such as that requiring inspection of tea and
another setting standards for butter and margarine, but it was only in 1906
that a comprehensive federal law was enacted. No doubt the long delay in
having such a law enacted owed much to the doctrine of 'State rights'.

The 1906 law defined the legal meaning of 'foods' and 'drugs' and
prohibited their introduction into inter-State commerce if misbranded or
adulterated. It also made illegal the use of many harmful chemicals in food.
The law marked a great step forward, but several weaknesses in the
legislation appeared in subsequent years. Another milestone was the
Shirley Amendment of 1912, which prohibited the labelling of medicines
with false therapeutic claims. In the 1930s public concern grew once more
about widespread sales of adulterated foods and spurious drugs. A
powerful stimulus to this disquiet was provided by Upton Sinclair's novel
The Jungle. Finally, in 1938, the much more effective and better designed
Federal Food, Drug and Cosmetic Act was passed by Congress. This law is
still the foundation of American food law. It provided, among other things,
that the Food and Drug Administration could set standards for the
composition of foods. It provided for ingredient labelling for those foods not
standardised. It prohibited the addition of poisonous substance to foods
and set safe limits on permissible levels of toxic substances which could not
be totally avoided. Some of the provisions of the Act had been adopted
from previously enacted State regulations. Such regulations had been in
force for many years, especially in those States which were large food
producers, and covered the manufacture, transport and sale of foods
within the State boundaries. Any food destined for sale outside the State,
whether within the USA or abroad, would come under the jurisdiction of
federal law.

Operation of the Food, Drug and Cosmetic Act is rather similar to that of the UK Food and Drugs Act. Like many other laws it is a broad 'enabling law' and statutory regulations are proposed by the agencies responsible for its implementation. After publication of an official announcement of proposed regulations in the federal register, there is a specified length of time during which comments may be made by interested parties, both US citizens and others. Several drafts of the proposal may be produced in the light of these comments and a hearing may be called if necessary. Only then are final decisions made by the Food and Drug Administration. The implementing regulations are published in the *Code of Federal Regulations*, Title 21, 'Food and Drugs'.

The 1938 law was not long in operation before amendment was needed to cover unforeseen loopholes and allow for new developments in science. In addition, it began to be accepted by many that there was a serious weakness in the absence of a requirement that the safety of food additives should be demonstrated before they were released on the market. Under the law any chemical could be used until the government proved in court that it was harmful, and the burden of proof was entirely on the government. In 1949 a select committee of the House of Representatives, the Delaney Committee, was appointed to investigate the use of chemicals in foods and cosmetics. Out of their deliberations came several amendments to the Act. The Committee adopted the principle that the safety of ingredients should be determined before the public is exposed to them. This fundamental change has been described as one of the most important advances in social legislation in the USA during the past century. 'It made the manufacturer responsible for establishing the safety of his products and the government responsible for evaluating the evidence submitted. The burden of proof was shifted from government to manufacturer.'

There have, of course, been many criticisms made of such mandatory pre-marketing clearance. A subsequent amendment known as the Delaney Clause (Food Additives Amendment of 1958, PL 85–929, and Colour Additive Amendments of 1960, PL 86–618) has been especially singled out as a target for lobbies seeking to weaken the safety requirements of the Act. The Clause banned from use as a deliberate additive in commercially formulated foods any substance which was found by appropriate test 'to induce cancer in man or animal'. Its opponents called the Clause unscientific, in that it banned compounds without considering their 'no effect' levels in diets; and wasteful, in that it forbade balancing benefits of use against risks. In fact, the Clause did not deal with trace doses of compounds that might be consumed by individuals, but only with batch amounts intentionally mixed with food for consumption by whole populations; and it did not foreclose uses beneficial to health, because it did not interfere with the sale of the same compounds as drugs. The Delaney Clause was seldom invoked to ban an additive, because the FDA preferred to curb suspected carcinogens by finding that they did not meet the more general requirements of the Act.[12] Its indirect influence, however, was great. In principle, it placed a growing list of suspected carcinogenic compounds in the same class of cumulative, chronic health hazards as redioactive isotopes, and subjected their use to similarly stringent restrictions.[13]

It may be useful here to compare the US system of control of food additives with the practice in some other countries. Selenium, of which we will have considerably more to say later, provides a good example of the application of such controls. As we shall see, selenium is an essential nutrient and as such is used as a supplement in some livestock feeds. The element can also cause acute toxicity and while dose–response relations for some animals are known, the hazards for humans are not yet completely elucidated. Because of such uncertainties, practice of control of the sale of selenium varies between different countries. The element is sold over the counter in New Zealand as a component of pre-mixed animal feed supplements;[14] in Sweden it is available only if dispensed by prescription from veterinary surgeons; and in the USA it may not be introduced into feeds as a chemical additive, but it is provided by supplementing feeds with alfalfa grown on naturally seleniferous western soils.[15] Each of the three countries appears to have found a cost-effective way of balancing the benefits of selenium against its hazards, even though the method used in each place may have been ineffective or unworkable in the other two.

INTERNATIONAL STANDARDISATION AND HARMONISATION OF FOOD LAWS

The history of the development of food laws in many of the other technologically advanced nations is very similar to that of the UK and of the USA. The change from home production of basic foodstuffs to dependence on shops and markets for supplies resulting from the Industrial Revolution, and increased urbanisation of the nineteenth century, led to the passing of many specific and, later, comprehensive food laws in Germany and other countries in Continental Europe. The major principles underlying all such food legislation were the protection of health and the prevention of fraud. Though sharing a common foundation, national differences in outlook and diversity of economic interest led to considerable variation between the actual laws passed in the different countries. Thus today even within the EEC there is a very great variety of laws and regulations governing the composition, labelling, handling and other aspects of food. As the world grows smaller in terms of travel and communication and international trade, the need for reduction of barriers to commerce grows. Pressure has been building up for harmonisation of national legislation relating to foodstuffs and even for the development of universally applicable standards for foods. The United Nations, through two of its agencies—the Food and Agricultural Organisation and the World Health Organisation—as well as, on a smaller scale, the EEC, have been taking steps to answer this need.

EEC Regulations

It is clear that the complexity and variety of regulations governing the composition and production of food in different countries serve as an obstacle to trade between nations. When the EEC was set up under the Treaty of Rome in 1957, one of its chief aims was stated to be the facilitation of trade between its member States. In spite of the fundamental relationship between the food laws in the different nations of Europe, the intricacies of legislation in each of them made freedom of movement of goods and expansion of inter-State trade in food difficult. Consequently, the Community instituted a programme of harmonisation which involves the formulation of legislation which will be generally applicable within the EEC but at the same time take account of the interests of individual member States. Such harmonisation of member States' food laws is not an easy task. Progress is slow since States are often unwilling to give up their own legislation and adopt that of others. However, some progress has been made.

As might be expected, the machinery for dealing with such EEC food legislation is complex—far more so than that involved in the development of similar laws in any one nation. Different bodies are involved: the Assembly or European Parliament, which plays a consultative role mainly; the Council of Ministers, consisting of one minister from each member State, which acts as the major decision-making body, and finally the chief administrative and executive organ of the EEC, which is the Commission. Its members are appointed by mutual agreement of governments of the member States. The Commission draws up rules and regulations for submission to the Council. Both Commission and Council are assisted by a number of advisory and consultative committees. By provision of the Treaty of Rome harmonisation of the food laws is achieved by regulations, which are laws binding in every respect and replacing equivalent domestic legislation, and by directives, which bind member States as to the final result to be achieved, but leave them free to choose the method in attaining this end. Such EEC legislation must pass through a number of stages before it comes into force. Initially, a draft is drawn up by the Commission in consultation with government experts from the member States and representatives of the industry and of consumer interests. The proposed legislation is then submitted to the Council and is also considered by the European Parliament and by the Economic and Social Committee. Finally the legislation, if acceptable to all, is promulgated by the Council. Such painstaking and slow-moving procedures lead to protracted delays in achieving the aim of harmonisation. As a consequence only a relatively small amount of legislation has been passed in final form up to the present. Much more is still in draft form. The text of EEC legislation is published in the *Official Journal of the European Community*, which appears daily in each of the European languages at Brussels. It should be noted that even though directives are agreed by the Council of Ministers they do not immediately become law, in member States. In Britain, for instance, they can only be implemented when a regulation is made under the Food and Drugs Act.

An example of an EEC directive of relevance here is that published on 4 May 1976, No 76/462. It deals with general criteria of purity of preservatives in foodstuffs and directs that 'they shall not contain toxicologically dangerous amounts of any element, in particular heavy metals; they shall not contain > 3 mg/kg of As or > 10 mg/kg of Pb; and they shall not contain > 50 mg/kg of Cu and Zn taken together, of which the Zn content must in no case exceed 25 mg/kg'. The potential for legal squabbles of the expression 'toxicologically dangerous amounts', unspecified except for arsenic, lead, copper and zinc, should be noted.

The Codex Alimentarius Commission
The establishment of economic groupings such as the EEC and similar organisations in Africa and Latin America, which are aimed primarily at creating common markets among member States, has stimulated interest not only within the groupings, but also outside them in the need to remove economic as well as non-economic barriers to international trade in foodstuffs. Practical interest in such moves has been shown by member governments of the FAO and of the WHO. Initially in 1958 an FAO

committee of experts collaborated with the International Dairy Federation to produce a Code of Principles on Milk and Milk Products. In the same year a body known as the Codex Alimentarius Europaeus was set up by the International Commission on Agricultural Industries and the Permanent Bureau of Analytical Chemistry. The aim of the body was the elaboration of internationally acceptable food standards. In 1961 the work of the European Codex was, by mutual agreement, taken over by the FAO and the WHO. A joint FAO/WHO Food Standards Programme was established and a joint subsidiary body known as the Codex Alimentarius Commission set up.

The objectives of the Commission are to develop international food standards and to publish these standards as the *Codex Alimentarius*. In line with these objectives a great deal of work has been done dealing with compositional, labelling, additive, contaminant, pesticide residue, hygienic, sampling and analytical aspects of foods. In collaboration with member governments, international agencies and other specialist bodies and following a detailed and prolonged procedure which allows adequate consultation with all interested parties, a number of standards for various foodstuffs have been proposed by the Commission. The standards are all presented in uniform manner and are intended to be adopted internationally. The basic aims of all the standards are to protect the health of consumers and to ensure fair practice in the food trade. In addition the Codex produces provisions of an advisory nature in the form of codes of practice and guidelines in the interests of international harmonisation of food laws. The Commission sets out to secure international agreement on the substance of its food standards and invites governments to accept them in various specified ways, either fully or with minor deviations, or only after a target date is reached. This method leaves governments free to follow their own national procedures and to advise the Commission on how the standards are to be implemented locally. At the present time considerable interest is being shown by many nations in the activities of the Codex Alimentarius Commission. Its views have, for instance, been taken into consideration in the UK in the preparation of legislation on food additives. Expert committees on food in other nations also draw on it for information and advice.

METALS IN FOOD LEGISLATION

Few countries have specific regulations relating to more than a few metals in foods. Some have general laws and directives covering all metals considered both as intentional and as accidental additives in food. Others do not legislate in this way, but only consider the metals in so far as they are of particular significance in certain foodstuffs for which specific regulations exist. For example, a country or an international authority such as the EEC, in setting standards for dried milk powder, may stipulate a maximum limit of so many milligrams per kilogram of a certain heavy metal in the product. Thus it is not easy to draw up a summary of laws and regulations relating to metals in food without either listing a considerable number of foodstuffs or entering into considerable detail, except for a small number of countries.

The UK is one nation which, because of the manner in which its food laws are formulated, has been able to produce a brief summary of legislation relating to metals in food. An officially produced *Summary of Regulations and Recommendations for Heavy Metals for the United Kingdom* is available from the Ministry of Agriculture, Fisheries and Food. These statutory and recommended limits are to a large extent due to the work of the Metallic Contamination Subcommittee, which was appointed by the Food Standards Committee in 1948. Largely as a result of the Committee's recommendations which have been published in official reports, statutory limits for lead and also for arsenic now apply to almost all foods. These limits are in addition to statutory maxima for certain metals in a small number of specific foods which had been promulgated previously in Food Standards Orders. As yet no statutory limits for other metals have been laid down, but there are recommended limits for copper, zinc and tin. These recommended limits are used as guidelines by enforcement authorities under the Food and Drugs Act 1955. All food must, of course, conform to the general provisions of this Act.

The UK *Regulations and Recommendations* may be summarised as follows:[16]

Lead: 2 mg/kg maximum permitted level in food; to be reduced to 1 mg/kg, with certain limits for specified foods.

Mercury: no statutory limits, but foods containing levels which are unacceptable in the country of origin are not to be permitted entry to the UK.

Cadmium: no statutory limits, but it is recommended that periodic analyses of certain foods should be carried out with the possibility of limits being imposed if necessary.

Arsenic: a maximum limit of 1 mg/kg for foods in general, with certain limits for specified foods.

Antimony: no general specified or recommended limits, except in the case of food colouring, where 100 mg/kg is laid down.

Copper: the general recommended limits are 2 mg/kg for ready-to-drink beverages and 20 mg/kg for other foods. There are no statutory limits. In addition there are special limits for specified foods and beverages.

Tin: the recommended limit for canned food is 250 mg/kg. No other limits are laid down.

Zinc: recommended limits are 5 mg/kg in ready-to-drink beverages and 50 mg/kg in other foods. There are also special limits for a few specified foods. A useful summary of these regulations is given in Pearson.[17]

Many of the smaller countries do not have detailed food regulations concerning standards of composition, additives, etc., though they normally have a general act which requires food to be of good quality. Such countries either accept foodstuffs complying with the regulations of the country of origin or with the regulations of a particular country. Thus, for example, Zambia and Tanzania accept foodstuffs complying with UK regulations, while Zaïre accepts those which meet the regulations of Belgium; Togo takes those of France. On the other hand, Sudan and Lebanon accept foodstuffs in general which meet the regulations that apply in the country of origin.

A number of other technologically developed countries besides Britain have produced summaries of regulations concerning the permitted limits of metallic contamination in foods. Many of the limits prescribed are very similar to the levels recommended or specified in the UK. The following summary is based on Lenane,[18] but up-to-date government regulations should be consulted for complete details.

South Africa
The following general regulations concerning trace metals in food apply:

Foodstuff	Arsenic	Copper	Lead	Zinc
	(mg/litre or mg/kg)			
Most non-alcoholic beverages	0.2	5.0	1.0	5.0
Other foodstuffs	1.0	20.0	5.0	50.0

with variations for certain specified foodstuffs such as gelatin (zinc 100 mg/kg) and solid pectin (arsenic 5.0, copper 300.0 and lead 50.0 mg/kg). A maximum level of 250 mg/kg of tin is set for all foodstuffs.

Canada
A similar short list of permitted levels of certain metals applies, covering beverages and selected foodstuffs. Examples of these permitted levels are as follows:

Foodstuff	Arsenic	Copper	Lead	Zinc
	(mg/litre or mg/kg)			
Most beverages as consumed	0.1	2.0	0.2	5.0
Fresh vegetables	1.0	50.0	2.0	50.0
Gelatin	2.0	30.0	7.0	100.0
Marine and freshwater animal products	3.0	50.0	10.0	50.0

Australia
This country has a more extensive coverage of individual metals and, as well as general limits for all foods, Australia has special regulations for certain specified foodstuffs. In addition, a distinction is drawn between levels of metal contamination in food packed in tin foil and tin plate and levels in food packed in other materials or unpacked. The general regulations, which are being reviewed at the present time, may be summarised as follows:

Foodstuff	Arsenic	Lead	Zinc	Antimony	Copper	Selenium	Tin
	(mg/litre or mg/kg)						
Beverages	0.15	0.2	5.0	0.15	5.0		
Other foods	1.5	2.0	40.0	1.5	30.0	2.0	40.0

with some exceptions which include the following:

Gelatin: arsenic 4.0 mg kg. lead 10.0 mg kg and zinc 100 mg kg.
Fish in tin plate: arsenic 1.5 mg kg, lead 5.5 mg kg and tin 250 mg kg.
Dried fruit: arsenic 4.0 mg/kg and lead 15.0 mg/kg.
Food packed in tin foil or tin plate may have a maximum of 250 mg/kg of tin.

Any other metal besides those listed above, as well as aluminium, calcium, iron lithium, magnesium, manganese, mercury and potassium, is restricted to a maximum level of 0.15 mg/kg in beverages and 5.5 mg/kg in solid foods.

New Zealand

Like Australia, New Zealand has specific regulations relating to a number of metals which have not yet been incorporated into UK or US food laws. The general regulations are as follows:

Foodstuff	*Arsenic*	*Copper*	*Lead*	*Zinc*	*Antimony*
	(mg/litre or mg/kg)				
All beverages	0.2	2.0	0.2	5.0	0.15
Other foods	1.0	30.0	2.0	40.0	1.0
except					
Food in cans or foil					
(with some exceptions,					
including beverages)	2.0	30.0	4.0	40.0	
Gelatin	2.5	30.0	7.0	100.0	
Vegetables	1.0	50.0	2.5	40.0	

In addition, New Zealand has set the following maximum levels for specific metals in various foods:

Cadmium: any food, 1.0 mg/kg.
Mercury: fruit and vegetables, 0.03 mg/kg; fish, 0.5 mg/kg.
Selenium: any food, 2.0 mg/kg.
Tin: generally, 40.0 mg/kg; but tin in canned tomato or asparagus: 250.0 mg/kg.

Unfortunately, it is not possible to make a neat summary such as those given above of US regulations and standards for metals in food. There is as yet no law which regulates a trace metal directly, though one is proposed for levels of mercury in fish and shellfish. Otherwise, information on metals must be obtained under the heading of different foodstuffs or animal feed or in standards for drinking-water. Standards which have been adopted by the US Congress are published in the *Congressional Record*. Administrative departments such as the Food and Drug Administration publish in the *Code of Federal Regulations*. Both volumes are published daily. In addition, acceptable standards for many food additives and other substances classified as GRAS ('generally recognised as safe') and accepted by the Commissioner of the Food and Drug Administration as 'food grade' are listed in the *Food Chemicals Codex*.[19] This is prepared by the Committee on Specifications of the Food Chemicals Codex of the National

Research Council and is printed by the National Academy of Science, Washington, DC. The latest (second) edition was published as a single volume in 1972, with two small updating supplements in 1974 and 1975. The volume gives analytical methods for many compounds, including trace elements, and also lists compounds in alphabetical order giving permissible levels of impurities, including trace elements, for each.

'The policy of the *Codex* with regard to trace impurities, including metals, is to set maximum limits for trace impurities wherever they are deemed to be important for a particular chemical, at levels consistent with safety and good manufacturing practice. The maximum limits for heavy metals shall be 40 parts per million, for lead 10 parts per million, and for arsenic 3 parts per million, except in instances where higher levels cannot be avoided (under conditions of good manufacturing practice). Where a heavy metals limit of 10 parts per million can be established, a separate limit for lead need not be specified'.[19]

The meaning of the term 'heavy metals' is explained later as 'common metallic impurities that are coloured by hydrogen sulphide (Ag, As, Bi, Cd, Cu, Hg, Pb, Sb, Sn) under the conditions specified'. The prescribed heavy metals tests involve precipitation of the metals from prepared samples as their sulphides by the addition of hydrogen sulphide solution. In addition to limits for such 'heavy metals', permissible amounts of selenium are also included. The *Codex* does not have the force of law, but has quasi-legal status as establishing food-grade quality. The *Food Chemicals Codex* specifications have also been adopted, under certain conditions, by the Food and Drug Directorate of Canada (29 January 1970) and by the Food Additives and Contaminants Committee of the UK Ministry of Agriculture, Fisheries and Food (1968).

Using the *Codex*, as well as the *Code of Federal Rules and Regulations*, it is possible to outline US regulations relating to trace metals in food as follows:

1. *Naturally occurring trace metals picked up by crops from soil*: no regulations.
2. *Trace metals entering food through pesticides, herbicides and other agrochemicals*: specifications exist for levels of each of several elements in pesticides and there are also limitations for pesticide residues in food. Thus trace metals in human food are indirectly limited. An example of this (*Code of Federal Regulations*, Title 21, Part 193.460) relates to a zinc–manganese co-ordination compound containing 20 per cent manganese, 2.5 per cent zinc and 77.5 per cent

ethylenebisdithiocarbamate. The residue permitted in certain foods is:

raisins:	28 ppm
bran of barley, oats, rye and wheat:	20 ppm
flour of barley, oats, rye and wheat:	1 ppm.

3. *Trace metals in drinking-water*: a recently passed federal law sets the following maxima:

cadmium	0.01 mg/litre	mercury	0.012 mg/litre
arsenic	0.05 mg/litre	barium	1.0 mg/litre
chromium	0.05 mg/litre	lead	0.05 mg/litre
copper	0.1 mg/litre	silver	0.05 mg/litre
selenium	0.01 mg/litre		

4. *Trace metals in animal feed*: maximum limits of selenium in chicken and swine feed have been set and there are proposals for levels in cattle feed.

5. *Trace metals in food additives*: there are various regulations, including that prohibiting the use of cobalt salts as a foam stabiliser for fermented beverages (*Federal Code of Regulations*, Title 21, Part 189.120). In addition, there is a long list of additives classified as GRAS, such as the following for gum ghatti:

arsenic:	not more than 3 ppm
heavy metals as lead:	not more than 40 ppm
lead:	not more than 10 ppm.

6. *Proposed standards for trace metals in food*: a recommendation that mercury in fish be limited to 0.5 mg/kg has been withdrawn and a 1.0 mg/kg tolerance level is now approved (Federal Register, 19 January 1979).

7. *Standards for trace metals in packaging*: the *Food Chemicals Codex* lists many materials in packaging which are subject to restrictions to prevent leaching of toxic substances into food.

8. *Trace metals in sewage sludge*: the Solid Waste Disposal Act includes sludge management practices. Sludge will have to be analysed for cadmium and other toxic substances before it is disposed of by land application. This Act has been substantially amended by the Resource Conservation and Recovery Act 1976.

9. *State Regulations*: these also exist for substances not included in federal legislation. Thus New York State has a legal standard for lead and cadmium in pottery.

CONCLUSION

It should be evident from these summaries of legislation from some of the English-speaking countries and, especially, from the points taken from some of the pertinent US documents, that consultation of the original laws and statutory regulations cannot be omitted by anyone who wants the full picture. There is growing pressure from consumer groups, environmentalists and many other concerned persons for further legislation in most countries to cover food contamination, pollutants and many other kindred matters. In the USA, for instance, there is a proposal for a uniform classification of hazardous substances in food according to a standard assessment of risks. In Germany, France and other non-English-speaking countries detailed legislation in this area is in the making if not already on the statute books. It is to be hoped that the plethora of new laws concerning metal and other contamination in food will not add further to the complexity and the diversity that already exists. If the nations of the world can only wait and pressure groups control their impatience, the United Nations through its joint WHO/FAO Committee should eventually produce a universally applicable and acceptable Codex covering food standards and legislating for permissible levels for metallic contaminants.

4

Food Quality—Analysis of Food

There are many textbooks and manuals available to the scientist who wants to analyse foodstuffs for their metal contents. Up-to-date volumes such as that by Pearson[1] devote chapters to the metals and also give useful information on more general analytical procedures and methods of identification and quantification of other food components. Numerous articles have appeared in the *Analyst*, the *Journal of the Association of Official Analytical Chemists* and elsewhere. For legally acceptable and officially recommended methods the scientist should consult the *Official Methods of Analysis* produced by the Association of Official Analytical Chemists, Washington, DC. This is now in its twelfth edition and is updated as the need arises. It is not intended here to attempt to offer a substitute for these works but rather to gather into one section a brief treatment of methods of analysis of metals in food, to draw attention where necessary to particular problems and in general to provide a readily consultable outline of the topic. This is not primarily intended for the professional analyst or for the skilled laboratory expert, but rather for that breed of scientist more often met than we sometimes are willing to admit, who has little more than an undergraduate foundation in analytical methods but needs now in the course of his investigations to analyse food and other biological samples for their metal components.

GENERAL METHODS OF ANALYSIS: SAMPLING

Obtaining a Representative Sample

Foodstuffs are seldom homogeneous in nature. Even bulk liquids can be stratified into layers of different concentrations. Thus if it is desired to apply analytical results obtained on a small sample or even a number of small samples to a large bulk, these samples should be representative. Sampling must be carefully carried out in order to minimise errors due to the inherent inhomogeneity of the bulk foodstuff. The same approach must be adopted even with less bulky samples when, as is often the case with animal and plant tissues, the nature of the material is naturally heterogeneous. In analysing leafy vegetables, for instance, it would be misleading to use samples which were largely made up of fine leaf fragments because all the midribs had been removed by sieving. In such cases the total sample will have to be homogenised and all portions included in the samples analysed.

Bulk material is normally selected for analysis either by random or by representative sampling. When random sampling methods are used, portions are taken in such a way as to ensure that every part of the material has an equal chance of appearing in the sample. It is important to avoid bias due to personal preferences of any sort. This can be done by numbering the containers and then using a table of random numbers to control selection by number, irrespective of external appearance or any other extraneous factor. Various sources for random numbers are available but one of the most widely used is Table XXXIII in Fisher and Yates's *Statistical Tables*.[2] Subsequent subsampling may be necessary by quartering or other method if the containers selected by use of the random tables are large, to ensure that representative portions are selected in each case.

Large bulk containers may have to be subjected to representative or stratified sampling. Here samples are taken in a systematic way so that each portion selected represents a corresponding portion of the bulk. The container is considered to be divided into different sections or strata and samples are removed from each section. The sizes of the samples should correspond to the relative proportions of the imaginary sections of the bulk. With bulk liquids or even with fine powders, thorough mixing by slow end-over-end rotation is generally preferable to representative sampling.

Grinding of Samples

Before being subjected to analysis, foodstuffs selected in the above or in any other manner will normally require further preparation. Few foodstuffs are absolutely homogeneous. They will consist of different portions of plant and animal tissue, unevenly distributed, and will require mechanical treatment to produce uniform samples. Dry foods can be chopped by hand and then mechanically ground using a ball or other grinder or a suitable blender. Further reduction and mixing in a mortar may be necessary. There is considerable danger here of contamination by trace metals. Absolute cleanliness of the chopping and grinding equipment is essential and surfaces should be of high-quality stainless steel. Even then contamination, especially with iron, chromium and nickel, is possible and checks should be made to see that misleading analytical results are not obtained in this manner. Often the ground-up food samples will need to be sieved for subsequent analysis. It may be important to specify the sieve mesh used and here again care must be taken not to contaminate the samples, especially with copper if brass screens are used. A fine nylon mesh has been found suitable by the author for avoiding this type of contamination.

Foods such as meat and meat products, which consist of several types of different tissue held together by connective tissue in a moist, mixed medium, are among the most difficult to sample. These foodstuffs should be thoroughly minced and then ground with a pestle and mortar before sampling. Less resistant foodstuffs such as fish or semi-liquid foods containing lumps of solid may be homogenised using a high-speed blender. With all these methods care must be taken not to contaminate the samples with trace metals from the mechanical parts, and thorough washing between use with different samples is essential. Where it is possible to use it, the Colworth Stomacher, in which food is contained in a plastic bag and never comes into contact with the metal parts, has proved to be an excellent means of blending samples.

Drying
Tables of food composition usually relate to fresh food or to portions as served, and these are the values often of most interest to nutritionists and dietitians. Similarly, statutory regulations and recommended levels of intake are usually based on wet or fresh weight. Thus analyses are frequently carried out on samples which have not been subjected to any treatment other than homogenisation or grinding. If such samples are not to be analysed immediately after selection they must be stored in suitable containers to prevent moisture loss or other changes that might affect analytical results. The use of heat-sealed plastic bags of a suitable size and storage at low temperature is a convenient way of ensuring retention of sample quality.

It is often desirable to perform analyses on dried samples, especially if the moisture content of the original material is variable. In many cases samples are most conveniently dried to constant weight in a fan-assisted hot-air oven maintained at 70–100 °C. Overheating or charring must be avoided. A vacuum oven is useful for allowing the use of lower temperatures and speeding up the drying process. Freeze-drying, if available, is highly desirable. It reduces the moisture content of the food to a suitable level and at the same time produces a crumbly texture, which is very convenient in subsequent sampling and analysis. The samples, when sealed in a plastic bag or other container, may be stored after freeze-drying without refrigeration.

PREPARATION OF SAMPLES FOR ANALYSIS: DESTRUCTION OF ORGANIC MATTER

Few foodstuffs can be analysed directly. Most contain organic matter which would interfere with the analytical process. The few exceptions are some beverages, including water, to which it is possible to apply colorimetric and other analytical techniques directly. Atomic absorption spectrophotometry for the estimation of lead, iron, zinc and other metals in wines and spirits which had not been given any prior treatment has been used successfully by Meranger and Somers[3] and for beers and spirits by Reilly.[4] A collaborative study reported by Strunk and Andreasan[5] showed that copper could be determined with accuracy in a wide range of alcoholic beverages without prior treatment.

Generally, however, metals are present in foodstuffs in such a manner that they cannot be estimated unless the organic matter of the food is first removed. This is usually done by some form of oxidation, either by the use of oxidising acids in a wet digestion or by dry ashing in the presence of atmospheric oxygen. The choice of method depends on the metals to be determined as well as on the nature of the food to be analysed. Generally the method chosen is that which gives reliability and accuracy in the range of concentrations appropriate to the investigation. At the same time it should permit the work to be carried out with reasonable speed using the facilities available to the investigator. The method actually chosen is often a compromise, for of the many techniques and procedures reported in the literature for the different metals, none will be able to give all the desirable qualities at the same time. A useful and informative comparative study of the different methods of oxidation of organic matter and their effects on subsequent recovery and estimation of the various metals has been made by Gorsuch.[6] This report was used by the Metallic Impurities in Organic Matter Subcommittee of the Analytical Methods Committee of the Society for Analytical Chemistry to prepare a series of recommendations on methods of wet and dry oxidation suitable for different metals. The report[7] forms the basis of the following section.

Dry Ashing

This is a convenient method for oxidising organic matter. It is suitable, if performed with proper care, for most metals though not for mercury or arsenic. It allows the use of larger samples than are usual in other methods of oxidation, avoids the danger of contamination from added reagents, does not result in large reagent blanks and moreover does not require constant attention by the operator. It is usually carried out at temperatures between 400 and 600 °C, but temperature and length of combustion will depend on the metal being analysed and the need to avoid loss of metal resulting either from volatilisation (especially with copper, selenium, antimony, cadmium, arsenic, mercury and some other elements) or by combination with the material of the crucible. Excessive heating may also make certain metallic compounds, such as those of tin, insoluble. Some flour products give a dark melt in which carbon particles are trapped and will not burn. The following method, based on one of those recommended by the AMC, modified in the light of other reports such as that of Crosby[8] and the *AO AC Official Methods*[9] and Gorsuch[6], has been found suitable for several common metals. The method is as follows:

Samples to be analysed should be spread in a thin layer over the bottom of a crucible. This should preferably be made either of platinum or of silica. Porcelain crucibles may cause high levels of metal retention. New, previously unused crucibles should not be used as these have been shown to retain more metal from samples than used crucibles.

The crucibles may be placed on a fire-clay triangle and heated over a low flame to bring about initial carbonisation of organic matter. They are then transferred to a temperature-controlled muffle furnace. Generally ignition at a temperature between 500 and 550 °C for about 8 hours has been found satisfactory except in the case of some very volatile metals when a maximum of 420 °C should be maintained overnight (8–12 hours).

The initial carbonisation over a low flame may be omitted if the samples are placed in a muffle furnace and maintained at a low temperature (100–150 °C) overnight, with a slight air draught for the first few hours. This prevents the build-up of carbon and waxy material on the crucible walls and lid.

The temperature may then be increased to a maximum of 550 °C. Oxidation is generally complete in about 8–12 hours, though longer ignition may be necessary with some foods.

The use of ashing aids is sometimes recommended to help decomposition of organic matter or to improve the recovery of metals, especially when the ash content of the sample is low. The aids are inorganic materials such as salts of various metals and some acids. They can be added at the start of ignition or after decomposition has been partially completed. The AMC recommendations give examples of several ashing aids suitable for low-ash material from which recovery of metals may be difficult.

Gorsuch considers that the use of ashing aids is not necessary except when the total ash of the sample is low. He notes that the use of nitric acid produces a clean ash readily soluble in dilute hydrochloric acid, but unless it is applied only in the final stages of ignition it may result in deflagration and considerable loss of metal. Though the use of sulphuric acid tends to slow down oxidation of organic matter it does have the effect of reducing losses by volatilisation.

Dry Ashing in an Oxygen Flask

This has been used for very small samples by some workers.[10] The sample is held in a platinum holder in a stoppered, oxygen-filled flask. The flask contains a suitable liquid for absorbing the products of combustion. Ignition is achieved by electrical excitation and takes only a few minutes. However, while attractive from the point of view of speed, the very small size of the samples makes it at present unsuitable for routine food analysis.

Wet Ashing

As Gorsuch has remarked with regard to lead, the choice of a method for destroying organic matter before determining trace amounts of metals is largely a personal one. Nevertheless the applicability of wet oxidation to a wide variety of materials, the relative rapidity with which it can be performed and the considerable superiority with regard to recovery of metals resulting from low volatilisation and limited retention loss, make it the most attractive method for many investigations. However, it has some disadvantages compared with dry ashing. In particular only small samples can normally be handled, using relatively large volumes of reagents. This can lead to high blank values. In addition, it is a potentially hazardous method and requires constant supervision.

Wet digestions are usually carried out in Kjeldahl flasks, preferably with some means of at least partial reflux of the hot acids. A suitable apparatus, consisting of a flask fitted with an extension to the neck and a fume condenser, is illustrated in the AMC report. Gorsuch found that when Kjeldahl flasks were not available, 500 cm^3 conical flasks were suitable. The apparatus should be made of borosilicate glass or silica. Heating temperatures should be kept relatively low and liquid conditions maintained by topping up if necessary in order to prevent losses by volatilisation and retention on the surface of the flask. Oxidising agents used include nitric, sulphuric and perchloric acids and hydrogen peroxide. Nitric acid boils at a relatively low temperature (120 °C) and tends to evaporate before the reaction is complete, thus requiring topping up at frequent intervals. This can increase the blank values, but the acid has the advantage that excess can be removed easily by heating at the end of the digestion. In addition, most metals form soluble salts with nitric acid.

Nitric acid digestion is not always sufficient for complete destruction of all organic matrices. Sulphuric acid, which boils at 338 °C, is usually added to the mixture to bring about complete digestion. If used alone, sulphuric acid tends to char organic matter and this can result in losses of some metals. In addition, it tends to form insoluble compounds with alkaline earth elements. These compounds may absorb other elements such as lead and resist full extraction. This is a particular problem with foods such as milk products which are rich in calcium.

Mixtures containing perchloric acid are widely used in spite of the potential danger involved. Gorsuch found that a mixture of nitric and perchloric acids was an excellent oxidising medium. It was particularly useful for the recovery of such metals as lead which form insoluble sulphates. Only with mercury was any significant loss found. However, it is essential to take precautions to avoid explosions during use. The Society for Analytical Chemistry has published a code of practice[11] which should be consulted before using perchloric acid in oxidation mixtures.

Hydrogen peroxide can also be used as an oxidising agent. Its great advantage is that it decomposes to water on heating. However, it can also be responsible for explosions and must be used with care.

The AMC report gives details of a number of wet digestion procedures using nitric and sulphuric acids, with and without perchloric acid or hydrogen peroxide.

An interesting method in which digestion is carried out in a tall beaker covered with a watch-glass is described by Middleton and Stuckey.[12] It involves heating on a hotplate with acid mixtures to dryness, remoistening with nitric acid and repeating the process until a white ash is obtained. The method is particularly suitable for animal and fatty material and does not require the constant supervision of other wet digestion methods.

Wet Digestion with Nitric and Sulphuric Acids
A generally applicable wet digestion technique which has been found

convenient and reliable, and is based on the AMC methods, is as follows:

A 5 g sample is placed in a 100 cm^3 Kjeldahl flask and 10 cm^3 of concentrated nitric acid followed by an equal volume of distilled water are added. The flask is heated gently to boiling and heating continued until the volume has been reduced by half. The contents are allowed to cool and 10 cm^3 of concentrated sulphuric acid are added gradually. Heat is again applied and small quantities of concentrated nitric acid added as soon as blackening of the contents is observed. Heating must be kept moderate to prevent excessive charring; a small amount of nitric acid must be present throughout the digestion. The process is continued until the solution fails to darken on prolonged heating to fuming (5–10 min). The final solution will range in appearance from colourless to pale yellow, especially if much iron is present. There may also be a precipitate which is soluble on dilution. The solution is allowed to cool, about 5 cm^3 of water added and the mixture reheated gently to boiling, when the solution should fume. Dilution to a suitable volume for further steps of analysis may be made with distilled water after cooling.

Use of added perchloric acid or hydrogen peroxide. The addition of small volumes of perchloric acid or hydrogen peroxide, after the stage in the above method where prolonged heating fails to darken the mixture, is recommended by the AMC to speed up oxidation, reduce the amount of nitric acid required and clear persistent residual colour. However, precautions must be taken to prevent explosions. Perchloric acid cannot be used if the presence of chlorides will interfere with subsequent estimation of metals. It is also not recommended when atomic absorption spectrophotometry with electrothermal atomisation follows.

Wet Digestion without the Use of Sulphuric Acid
To avoid the use of sulphuric acid, which can interfere with the estimation of lead and other metals, especially in the presence of alkaline earth metals, the AMC recommends a method using nitric and perchloric acid only. It is of wide applicability and is especially suitable for the destruction of protein and carbohydrate material. It is not, however, suitable for very fatty samples. It can be used for most metals with the exception of mercury. It is a relatively safe method, since the absence of sulphuric acid reduces the operating temperatures. However, care must be taken not to allow the mixture to boil dry. The procedure is as follows:

A sample containing not more than 2 g of dry matter is placed in a

200 cm^3 Kjeldahl flask. 25 cm^3 of nitric acid, (sp. gr. 1.42) are added and the flask boiled gently for 30 min. On cooling, 15 cm^3 of perchloric acid (60 per cent w/w) are added and the mixture boiled very gently until the mixture is colourless, or almost so. Thick, white fumes should be visible at this stage. Continue boiling for an hour, with constant supervision to prevent drying out of the flask contents. The flask contents may be further diluted with water on cooling and made up to a suitable volume. Most perchlorates resulting from this digestion will be soluble in water.

As will be mentioned when the individual metals are considered, special digestion methods are necessary for mercury, arsenic and some other metals under particular conditions.

PREPARATION OF SOLUTIONS FOR SUBSEQUENT METAL ANALYSES

In the case of ash prepared by dry oxidation techniques as described above, it is sometimes sufficient to add a suitable volume of dilute hydrochloric acid to it to bring the metals into solution. However, a procedure based on the AOAC method[9] has been found to give better results in many cases.

The ash is moistened in the crucible with a minimum volume of hydrochloric acid (diluted 1 to 1 with distilled water). Care is taken not to allow any loss by spurting. Approximately 20 cm^3 of distilled water are now added and the crucible is placed on a steam bath and evaporated to near dryness. 20 cm^3 of 0.1N HCl are added and heating continued for a further 5 min. After cooling, samples may be made up to a volume suitable for the next steps of analysis by addition of 0.1N HCl. In some cases where there is still an insoluble residue, filtration may be necessary before proceeding.

Another dilute acid mixture often used to dissolve ash is HCl, HNO$_3$ and water in the proportions 2 : 1 : 3.

Frequently, digests prepared by wet oxidation can be used directly when suitably diluted for subsequent metal determination, especially if atomic absorption spectrophotometry is employed. However, interference effects from other substances present in the solution may cause problems, especially if determinations are to be carried out at levels of concentrations approaching the limits of detection.

Chelation and Solvent Extraction

These problems may be overcome by pre-treatment of the sample in order to concentrate the metal of particular interest with respect to competing substances. This may be achieved by chelation, followed by solvent extraction of the metal chelate into an organic phase.

Many different complexing agents are available, and methods using them are reported in the literature. They include ammonium pyrrolidinedithiocarbamate (APDC), diphenylthiocarbazone (dithiozone), nitrosophenylhydroxylamine (cupferron), quinolin-8-ol, oximes and dioximes, beta-ketones and others.

They vary in their selectivity and are usually very dependent on the pH at which extraction is carried out. Dithiozone, for instance, which is widely used for the extraction of metals for subsequent spectrophotometric analysis, forms complexes with about 20 metals. However, if pH is controlled its specificity is increased. While most of the metals are chelated under alkaline conditions, only copper, mercury and silver are extracted from strongly acidic solutions. Specificity can be further increased by the addition of other complexing agents such as cyanide and citrate to the solution before extraction. For the estimation of lead, for instance, the AMC proposed[13] that ammonium citrate as well as potassium cyanide be used and that the dithiozone chelate of lead be taken up by chloroform.

Besides chloroform, several other organic solvents are used, including carbon tetrachloride and benzene. Choice of solvent will usually depend on its suitability for subsequent stages of analysis.

Ion-exchange Resins

These are finding extensive use in removing interfering metals from solutions in preparation for final analysis. A recent development has resulted in the introduction of chelating ion-exchange resins in which ligands of various types are attached to solid supports. Their use allows considerable concentration enhancement. Baetz and Kenner[14] have used a column containing an iminodiacetate chelating resin (Chelex 100) to concentrate metals in a digest solution following oxidation. The solution was passed through the column and after elution metals were determined by atomic absorption spectrophotometry. Recoveries of around 97 per cent were achieved for cadmium, copper, manganese, nickel, lead and zinc in such foods as spinach, fish, apples, cereal products, milk and also in leaf and liver reference materials.

END-DETERMINATION METHODS OF METAL ANALYSIS

The actual method employed for the analysis of metals in samples prepared as described in the preceding section will depend to a large extent on the facilities available to the investigator. Where financial support is generous and emphasis is on exactitude and on the development of ever finer techniques, then elaborate and costly methods such as X-ray fluorescence spectrometry or, when a source of atomic radiation is at hand, neutron activation analysis, may be the preferred methods of analysis. But for many research workers and for the scientists in a commercial quality-control laboratory, the choice will, of necessity, be the less costly spectrophotometry or, increasingly today, atomic absorption spectrophotometry. In fact, choice of end-determination methods will more often depend on laboratory space available, the number of samples to be analysed, the availability of skilled technical assistance, funding and similar considerations than on strictly scientific parameters. While we will look briefly in this section at most of the methods available today, irrespective of cost or complexity, our main interest will be in those methods most readily available and not beyond the competence of a scientist with at least moderate laboratory skills and the patience and intelligence to follow instructions from the specialist literature and the pertinent handbooks.

Spectrophotometry
A great number of analytical techniques are available to the food scientist today. For more than a century analytical chemists have been perfecting methods for the analysis of inorganic elements and compounds and many of their techniques have been applied with success to problems of food chemistry. Up to the 1960s spectrophotometric and polarographic methods were normally the most commonly used for metal analyses in food. But with the development, first, of flame emission and, most recently, of atomic absorption spectrophotometry, there has been a decline in the popularity of the earlier methods. Nevertheless, in laboratories where there is no need for routine metal analyses and where the capital investment in an atomic absorption spectrophotometer is not considered justified, or in field or other conditions where such an instrument cannot be used, chemical methods relying on spectrophotometric measurements are still widely used. Indeed, in his *Instrumental Methods of Food Analysis*,[15] MacLeod has listed a considerable number of recent applications spectrophotometry to metal analysis in food.

Such methods will later be discussed for individual metals. In their favour is their simplicity and cheapness, as well as sensitivity in many cases. However, they are sometimes lacking in specificity. An excellent treatment of the whole question of spectrophotometric and colorimetric determination of metals up to the beginning of the atomic absorption era is given by Sandell.[16]

Polarographic Methods

Though it is a technique requiring greater skills than does spectrophotometry, and also more elaborate facilities, polarography was fairly widely used in the pre-atomic absorption days and is still frequently used, especially in the related technique of anodic-stripping voltametry. Both techniques rely on the fact that different metals require the application of different electrical potentials before they are deposited from solution on to a cathode. A characteristic half-wave potential results for each metal and can be used for identification, while the height of the wave gives a measure of the concentration. The technique is sensitive if used with care. It is especially useful for the simultaneous determination of several metals, particularly the heavy metals. Organic matter must be destroyed prior to analysis. Oxygen must also be removed by bubbling the solution with nitrogen during the estimation. Temperature must be controlled to within $\pm 0.5°C$ for accurate results. The principles involved are discussed by MacLeod.

Anodic-stripping Voltametry

This method is even more sensitive than the older technique. After the metal ions have been pre-concentrated from the solution and amalgamated into a stationary hanging mercury drop electrode, the process is reversed by using a potential more negative than the reduction potentials of the metals of interest. The metals are oxidised and stripped anodically, using a slowly increasing positive potential. The measured current recorded during the stripping step is a direct linear function of the bulk concentration of each metal. The technique has been shown to be as sensitive and reliable as atomic absorption spectrophotometry. A recent application which illustrates its usefulness was reported by Hundley and Warren,[17] who used the technique to measure levels of cadmium in total diet samples.

Atomic Spectroscopy

It has been known for a long time that when a metal or its compounds are introduced into a flame, an atomic vapour of the metal is produced and coloured light is seen. The colour will be characteristic of the metals involved. The great firework makers of the Baroque period in Europe and, no doubt, their predecessors in China, knew how to get a green flame from copper and yellow from common salt, and many more such pyrotechnic effects. This knowledge was first put to more scientific use in the early nineteenth century when Talbot began to study flame spectra, especially of strontium and lithium. It was by such means that Bunsen and Kirchoff in the 1860s discovered the existence of caesium and rubidium. In the present century, flame spectra have been used very effectively in many types of analysis, and have provided us with analytical tools which allow us to discover secrets of nature at a level of minuteness undreamt of by our Baroque and even our nineteenth-century predecessors.

In the exothermic reaction of two gases such as hydrogen and oxygen which constitutes a flame, the initial reaction involves the breaking of bonds in the gas molecules and is endothermic. As a result, free radicals are produced. Usually the different free radicals will combine together, possibly in new arrangements, with no release of energy. However, if the free radicals come into collision with another body, whether it is a compound or an element or simply a solid surface, some of the energy will be released. A certain amount of this energy will be retained by the system and cause dissociation of incoming gases to bring about a chain reaction. Collisions between these free radicals and the atoms of an element also produce free radicals from these atoms. If sufficient energy is available the new free radicals may be raised to an excited state in which the valence electrons of the metal atom are raised from their normal to a higher energy level. This transition is immediately followed by the emission of radiation characterisitc of the element involved as the excited electrons drop back to the original energy level and the atom returns to its ground state. This behaviour provides the basis for the analytical procedure of emission or flame photometry. Only a small number out of all the atoms of the metal will be excited, even in a hot flame. However, the unexcited remainder can be made to absorb radiation of their own specific resonance wavelength (in general the same wavelength they would emit if excited) from an external source and in this way reach the higher energy state. Thus if light of this wavelength is passed through a flame containing atoms of the element, part of the light will be absorbed and the absorption will be proportional to the density of the atoms in the flame. This phenomenon is the basis of atomic absorption spectrophotometry.

Both emission and atomic absorption spectrophotometry use basically the same equipment and many commercial models have dual functions. Essential requirements are: (1) a means of atomising the test solution—an energy source such as a flame system into which the sample solution is aspirated at a steady rate, of sufficiently high temperature to produce an atomic vapour of the element of interest; (2) a stable light source, emitting the sharp resonance line of the element to be determined (for atomic absorption spectrophotometry only); (3) optical and electronic equipment for selecting the particular energy wavelengths of interest and recording levels of emission or absorption of light. The equipment can be relatively simple, as was much of the early laboratory-built equipment, or fairly complicated, as are some of the commercial dual-beam, automatic instruments available today. In no case, however, is it so expensive as to be beyond the reach of most well-equipped analytical and research laboratories.

Instrumentation for Atomic Absorption Spectrophotometry

Atomic absorption spectrophotometry is probably the method of choice for most scientists who are not formally trained in the disciplines of analytical chemistry or instrument engineering. According to Delves, atomic spectroscopic methods are the most widely used techniques for trace metal analysis in clinical laboratories.[18] Since the method was first developed by Alan Walsh of the Australian CSIRO in the mid-1950s and instruments became commercially available in the early 1960s, atomic absorption has had a rate of growth probably more rapid than that of any other previously developed instrumental analytical technique. Ease of preparation of samples for analysis, accuracy, high degree of reproducibility and the wide range of more than 60 elements which can be determined by it, in concentration ranges from trace to macro quantities, have all contributed to the popularity of the technique. Atomic absorption spectrophotometry has, it must not be overlooked, certain disadvantages. It is not, in fact, a qualitative method and the identification of an element must be known before it can be analysed quantitatively. Moreover, the method is restricted largely to metals, though indirect methods for some non-metals have been developed. Nevertheless, it is undoubtedly a method of major interest to a great number of scientists concerned with metals in food, and for this reason we will spend some time here considering it in some detail.

Overall Design

A single-beam instrument consists of a flame atomiser, a hollow cathode lamp, a monochromator and a recording device. By cutting out the external light source an instrument can be converted from the absorption to the emission mode. Double-beam instruments are also available. These compensate for changes in the intensity of the energy emitted by the lamp (especially drift during the warming-up stage) but not for fluctuations in the burner system.

Hollow cathode lamps are designed to emit energy of a particular wavelength and of narrow width, capable of causing atoms of the element of interest in the sample to be raised from the ground to the excited state. This is done by making the cathode of, or at least lining it with, the metal. The cathode is usually in the shape of a cylinder about 1 cm in diameter and 1 cm deep. When a current flows between the anode and cathode, metal atoms are 'sputtered' from the cathode cup and collisions occur with the filler gas. A number of the metal atoms become excited and emit their characteristic radiation. The choice of filler gas depends on its emitted spectrum compared with the spectrum lines of the element of interest. For example, while neon is suitable for lead, iron and nickel, it produces a strong emission line near to the lines emitted by lithium and is therefore unsuitable for this metal.

Most lamps are run at low current, since otherwise they can become overheated and damaged. Lamps also have a tendency to self-absorb the radiation emitted and this tendency increases at higher current, thus reducing the sensitivity. Most hollow cathode lamps are designed for a single element. However, multi-element lamps are available, though some have only limited usefulness due to emission of adjacent lines by elements not being measured. A useful mixed-element lamp is, however, calcium–magnesium. It is particularly applicable to the study of biological samples containing these two elements. It eliminates the warm-up time for two separate lamps and allows quick change-over from one element to another. High-intensity hollow cathode lamps have been found to improve the sensitivity of atomic absorption spectrophotometry for a number of metals, including nickel.

Microwave-excited electrodeless tubes have also been used with a number of elements, such as arsenic and selenium. However, at the present time most of the needs of the investigator in the field of trace metals in food will be met by the standard single-element hollow cathode lamps.

Modulation of the lamp beam from a dc to an ac signal, while the flame signal remains dc and the detector is tuned to pick up only the lamp signal, is used to overcome the problem caused by the fact that elements in many cases tend to emit radiation at the same wavelength at which they absorb it. This means that when light emitted by the lamp is absorbed by the atoms of a metal in the flame and so the intensity of the emission is reduced, the atoms in the flame immediately emit radiation at the same wavelength, and the level of emission is once more increased. But by modulating the beam from the lamp and tuning the detector to this frequency, the emitted light is no longer detected and the full level of absorption of the lamp emission can be detected.

Modulation is usually achieved by using a mechanical chopper rotating at a controlled frequency to which the detector is tuned. The chopper blades allow through blocks of light alternating with darkness and this produces an alternating current in the detector.

The monochromator filters out other radiation lines which may be produced by the lamp, apart from that which the sample can absorb. It should be capable of using wide shutter slits (wide band passes) for elements such as zinc, arsenic, calcium and lead which have fairly simple spectra. The wider the band the more energy passes through the monochromator, and so precision and detection limits are improved. In most modern instruments quartz prisms or diffraction gratings are used, rather than simple filters for the monochromator.

The burner, or as it is often known, the atomiser, must be made from corrosion-resistant materials to survive high temperatures. Different designs are required for the different gas systems, but they must all have certain attributes: they must have good stability and high sensitivity; they must not clog when aspirating concentrated solutions and must at the same time be capable of taking dilute solutions; they must also be versatile, suitable for a wide range of elements. Two basic designs are widely used. The total consumption burner is more often employed in flame than in atomic absorption work. Fuel, oxidising gas and sample solution enter through separate channels but emerge together at the same ignition point. The sample enters the flame as a fine mist, sucked up by the venturi effect. This design avoids the danger of blow-back, in which unconsumed gases explode. It is versatile and can be used with several gas systems, but the outlet is easily clogged and the flame produced is small and not very suitable for atomic absorption spectrophotometry. In addition, the flame is turbulent, with high physical and electrical 'noise'.

The pre-mix burner has generally replaced the total consumption unit for atomic absorption. Fuel, oxidant and aspirated sample are mixed in a separate chamber before entering the burner. Large drops fall to the bottom of the chamber and only a fine spray gets into the flame. The burner is narrow with a long slit, producing a long, non-turbulent flame—excellent for absorption. It is not easily blocked but there is danger of blow-back. Provision is made in most modern instruments for containing this if it occurs.

Fuels and oxidants. Air–acetylene is the most commonly used gas mixture, producing a flame of about 2400 °C. Higher flame temperatures are required for refractory elements such as aluminium, barium, beryllium, titanium and vanadium. For these a nitrous oxide–acetylene mixture is used, giving a flame of 3200 °C. Because of the increased danger of blow-back with this mixutre a shorter burner made of stronger materials is used. Other mixtures such as nitrogen–hydrogen, air–hydrogen and argon–hydrogen are also used for special purposes. Most burner systems will have a number of controls which can be used to regulate angle and height.

In conventional atomic absorption spectrophotometry the sample must be in solution, and it is introduced into the flame as a cloud of fine droplets by nebulisation. Generally, solutions prepared as has been described above are suitable for nebulisation, provided they are free of particulate matter, are not viscous in nature and have a sufficiently low concentration of total solids (about 2 per cent) to prevent clogging of the atomiser and reduce various types of interference known to be related to the physical and chemical properties of solutions.

Various methods have been suggested in recent years for overcoming such problems resulting from the method used in these two types of burner for introducing the sample into the flame. One procedure uses the Delves microsampling cup.[18] In this method the sample is placed in a small container with a little hydrogen peroxide, and the container is then put directly into the flame. The Delves cup has been used effectively for estimation of lead in blood. The particular advantage of this and a number of similar methods is that samples require only minimal pre-treatment and they allow estimations to be performed on samples of microgram and microlitre quantities. However, mainly because of the small sample sizes, extreme care is necessary if contamination and other extraneous influences on analytical precision are to be avoided.

Flameless atomic absorption spectrophotometry using, for example, electrothermal atomisation techniques, has been developed in recent years in order to avoid problems associated with the flame. Cells made of silica and quartz or graphite tubes and rods electrically heated to about 3000 °C are used. Samples of the order of 100 μl or less are introduced into the cells by micropipette. A high concentration of absorbing atoms in a small well-defined volume is produced. The technique has resulted in considerable improvement in detection limits and also allows speeding up of de-terminations. Some commercial models have a graphite microfurnace programmed in steps to dry and ash the sample directly *in situ* before atomisation. However, matrix and other interference effects can cause problems and the overall cost of operating such a system can be high, as the cells need to be replaced frequently. Reports have been published on the application of electrothermal atomisation to atomic absorption spectro-photometry in the analysis of a variety of biological materials.[18] The method has been found to show satisfactory correlation with other techniques as regards precision. It was also more rapid and more sensitive than the flame technique. The method has been particularly useful in determining low levels of copper and zinc in microlitre samples of blood.[19]

Atomic Fluorescence Spectrophotometry
This technique appears to offer considerable advantages compared with conventional atomic absorption techniques. It is capable of lower limits of detection for several metals and allows for multi-element determination. However, the method is still in its infancy and several technical problems need to be solved before it can be considered a rival for the better established absorption method. Browner[20] has reviewed the technique in some detail and can be consulted for further information.

Flame Emission Spectrophotometry
Though lacking some of the advantages of the atomic technique, this technique is by no means discarded in today's analytical laboratories. It is still widely used, particularly for alkali metals. From the point of view of economy and simplicity, it has advantages over the absorption method. Though, as has been mentioned, many atomic absorption spectrophoto-meters provide the emission mode as an auxiliary facility, simple and inexpensive single-purpose flame photometers are available.

Samples are prepared as for atomic absorption spectrophotometry. Unlike atomic absorption, the signal obtained in emission is directly proportional to the concentration of atoms in the solution and is linear over a wide range. This allows measurements to be made over a wide concentration range, usually in the region 0.1–50.0 mg/litre. The same effects of viscosity and surface tension on nebulisation are met with as in atomic absorption. In addition, the flame technique is subject to other chemical and spectral interferences for which corrective procedures have to be taken. Thus there can be interference between several metals and calcium as well as between potassium and sodium, calcium and mag-nesium. Anions can also cause problems, particularly phosphate and sulphate. The use of an internal standard such as lithium is recommended for overcoming such problems in the estimation of sodium and potassium.

X-ray Fluorescence Spectrometry
It is unlikely that many workers except those in large commercial or

government laboratories will have access to some of the analytical techniques currently being developed for the estimation of metals in food. X-ray fluorescence spectrometry, for instance, is applicable to all elements with an atomic number greater than 11, but the apparatus is costly and requires the services of skilled technicians. The technique requires only limited preparation of samples, mainly in making pellets for excitation. However, matrix problems are severe and the technique has not yet been sufficiently developed to allow its use in trace metal analysis.

Neutron Activation Analysis
Another analytical technique beyond the reach of the majority of scientists but of very great interest is neutron activation analysis (NAA). Capital cost is high and access to a nuclear reactor is usually necessary. Where such facilities have been available, the method has been shown to be the most sensitive of all techniques for the determination of trace amounts of metals in biological systems. Moreover, it has the considerable advantage of permitting multi-element determination on very small samples and is non-destructive. The sample is bombarded by neutrons so that some of its atoms are converted into radioactive isotopes. Gamma radiation is emitted during the decay of these nuclides. The energy of the emitted radiation is characteristic of the element and this radiation can be measured by a gamma spectrometer. The effectiveness of the technique has been shown, for example, by the work of Clemente,[21] who used it to measure levels of cobalt, chromium, caesium, iron, mercury, nickel, rubidium, antimony, scandium, selenium and zinc in the environment as well as in food and in human tissues. A useful report[22] of the use of neutron activation analysis for trace metals in food by the US Food and Drug Administration has recently been published.

A portable NAA system which uses a plutonium 238–beryllium source has been found effective for *in vivo* determination of cadmium and some other metals in humans.[23]

Other techniques, such as mass spectrometry, also allow sensitive, simultaneous determination of a wide range of elements in foods. Gas chromatography of metal chelates has also been applied to the analysis of biological tissues. No doubt other techniques will be developed in future years and those presently used will be improved.

PART II
The Individual Metals

5

Lead

'PESTILENTIAL AND NOXIOUS METAL'

When Georgius Agricola published his *De Re Metallica*[1] in 1556 he was, it seems, in a state of mind which should not be unfamiliar to us today. He was a scientist, devoted to his chosen field of metallurgy. He wanted others also to appreciate the wonders of the metals and to value the contribution they had made to civilisation. But he found that the metals and those who worked with them were held by many in low esteem. The life of a miner or a metallurgist was considered to be one of 'sordid toil'. Metal workers were accused of destroying the countryside with their pits and mines, polluting the streams with waste and generally harming rather than helping mankind. Among the various metals so decried, lead was particularly singled out as 'a pestilential and noxious metal'.

Agricola wrote his book to refute these criticisms. He wanted to show that 'if we remove metals from the service of man, all means of protecting and sustaining health and more carefully preserving the course of life are done away with. If there were no metals, man would pass a horrible and wretched existence in the midst of wild beasts.' His beautifully produced treatise may not have refuted every one of the accusations made against the metals and those who worked them, but it did give his contemporaries, and us, a magnificent record of the mining and metallurgical techniques in use in sixteenth-century Europe.

Agricola should not really have been surprised that lead was so heavily criticised by his contemporaries. Men had been pointing out the dangers associated with the use of the metal and its products for at least 2000 years. The Greeks had even given the name of the metal, or at least that of its associated planet, Saturn, to a form of poisoning from which lead workers were known to suffer. Saturnism, or plumbism as it came to be known later, described the colic, sometimes accompanied by delirium and paralysis, that Hippocrates had noted in about 400 BC in men who worked with lead. In the time of Pliny, at the beginning of the Christian era, shipwrights who painted white lead (lead carbonate) as a preservative on hulls, wore cloths over their mouths to protect themselves from ill effects. Later writers also commented on the health hazards associated with the metal, but in spite of this poisonings continued for centuries even after Agricola's time. It was not until the nineteenth century that the study of occupational medicine was actively promoted, and detailed records of the hazards of certain industrial materials and practices began to be kept. The picture painted by the reports of such men as Ramazzini in Italy, Thackrah in England and Tanqueril des Planches in France, showed that nineteenth-century Europe desperately needed laws to control industrial practices which exposed workers to hazards from poisonous substances. With reluctance on the part of governments and industrialists, the findings of these pioneers of industrial hygiene were finally accepted and steps were taken to eliminate the evils. Lead figured prominently in early legislation in the UK, for example in the Factories (Prevention of Lead Poisoning) Act 1883 and in regulations forbidding the consumption of food and drink in sections of pottery works where lead glazes were employed. These were but the beginnings of a long series of Acts and regulations devoted to preventing lead poisoning. As the previous chapter on Food Legislation has shown, lead is also prominent in the laws of many other countries besides the UK. Indeed, it is probably the metal which has had more official consideration and about which more has been written with regard to food contamination than any other inorganic contaminant. For this reason it seems right that we should devote more space to lead in this section of the present study than to the other metal contaminants of food.

CHEMICAL AND PHYSICAL PROPERTIES

Lead is element number 82, lying in group IVB of the Periodic Table, with an atomic weight of 207.19. It is one of the heavier of the elements, with a density of 11.4. It is soft and bendable and can be beaten flat with a hammer and cut with a knife. When first cut, its surface is bright and mirror-like, but this soon tarnishes on exposure to moist air as a surface film of basic lead carbonate forms. A similar protective film forms when the metal is placed in 'hard' water containing carbonates. Lead melts at a relatively low temperature of 327.5 °C. Its boiling-point is 1725 °C. The metal is a poor conductor of heat and of electricity. Oxidation states are 0, + 2 and + 4. In inorganic compounds lead is usually in state + 2. Most of the salts of lead (II), lead oxides and lead sulphide are only slightly soluble in water, with the exception of lead acetate, lead chlorate and, to a lesser extent, lead chloride. Lead forms a number of compounds of technological importance. When heated in air, a yellow powdery monoxide, PbO (litharge), is formed. On further heating the monoxide is raised to a higher oxidation level, producing trilead tetroxide, Pb_3O_4 (red lead).

White lead or basic lead carbonate, $Pb_3(OH)_2(CO_3)_2$, is prepared commercially by the action of air, carbon dioxide and acetic acid vapour on metallic lead. Another coloured compound is yellow lead chromate, $PbCrO_4$.

Some of the organic compounds of lead are of great economic importance. Tetraethyl lead, $Pb(C_2H_5)_4$, is liquid at normal temperatures with a boiling-point of 200 °C, and has antipercussive properties. It is prepared by the reaction of ethyl chloride, C_2H_5Cl, with sodium–lead alloy. Tetramethyl lead, $Pb(CH_3)_4$, is also a liquid, with a boiling-point of 110 °C. Both these organolead compounds decompose at or just below their boiling-points.

Lead has an ability to form alloys with other metals. Some of these, such as solder which is made with tin, are of considerable economic importance. Formerly, pewter contained 20 per cent or more of lead alloyed with tin, but modern pewters contain little, if any, lead.

PRODUCTION AND USES

Lead is found, at least in small amounts, almost everywhere in the world. Soils normally contain between 2 and 200 mg/kg. The metal does not occur in a pure elemental state, but combined with other elements as salts. It is usually also associated with other metals, especially zinc, iron, cadmium and silver. Economically workable ore bodies occur in many parts of the world. The major lead-mining countries are the USA, USSR, Australia, Canada, Peru, Mexico, China, Yugoslavia and Bulgaria. The most common ores are galena (lead sulphide), cerussite (lead carbonate) and anglesite (lead sulphate). In 1975 world production of lead was approximately 3.5 million tonnes, a million tonnes more than the amount produced 10 years previously. World lead consumption is expected to be about 6 million tonnes by the year 2000.[2] The ease with which lead may be extracted from its ores probably accounted for its early exploitation by man. When, for example, galena is roasted on a charcoal fire it is partly oxidised at quite a low temperature and then proceeds, more or less by itself, to complete the conversion to a metallic state. The lead oxide reacts with the unconverted lead sulphide and releases the free metal. This is the system on which the original ore hearth furnace was based, a method used by the Romans 2000 years ago and still used, until a short time ago, in a few places. Today it has been replaced by the blast-furnace, in which reduction is achieved by carbon monoxide or producer gas made from coke. The technique with sulphide ore (3–8 per cent lead) includes concentration (to 50–70 per cent lead), sintering, in which the lead is oxidised, followed by reduction and then, finally, refining of the metal.

Not all the lead produced today comes from ore. Of the 5 million tonnes used annually, almost 1.5 million tonnes of 'secondary metal' are produced from scrap. In the UK and the USA about half of the refined lead produced comes from recycled material.[3]

The uses of lead fall into two main groups: first in metallic form, and second in chemical compounds. In Britain, two-thirds of the total consumption comes into the first group and is related to the metal's high resistance to corrosion. A large fraction of the lead compounds in the second group go into additives in petrol.

The largest use of lead throughout the world today is in lead batteries—electrical accumulators for cars, electrical vehicles, emergency lighting and other such uses. This accounts for 40 per cent of total world consumption. In the UK approximately 100 000 tonnes are used annually for battery plates. It has been estimated that almost all the lead used in batteries is recycled from scrap, and the turnover time is about four years.

Lead-sheathed cables form another major use. About 60 000 tonnes are used in this way annually in Britain. Formerly a great deal of the metal was used in the manufacture of lead pipes, but with the introduction of plastic piping this use decreases year by year. Similarly, lead sheeting is used less today than in former times for roofing. It is still extensively employed in chemical plant for handling sulphuric acid.

Lead finds a variety of other uses. It has traditionally been used for making bullets and other projectiles for guns and is still employed in great quantities for this purpose. In France alone, for instance, some 12 000 tonnes of lead were used in the manufacture of shot for gun cartridges in recent years.[4] Because, as we shall see, shot discharged at game can be a source of food contamination, efforts are being made today to replace lead in shot by other less toxic metals.

Lead solders of various composition are used extensively in a number of industries, especially in plumbing and electrical work. Their use for joining seams in cans may result in contamination in food. In the motor industry, lead and lead solders are used in car body manufacture, in bearings and in many of the uses already mentioned. In fact, more than 50 per cent of the total world production of lead is for the motor industry, either in engine and body manufacture and accumulators or in fuel additives. Lead alloys are used in considerable quantity as printing metals, though normally the same metal is used over and over again as type-face is melted down and recast.

The use of lead compounds, both organic and inorganic, is extensive and continuously increasing. A certain amount of lead monoxide is still used today to manufacture white paint, but it is now largely restricted to paints used for priming or external work. Formerly much greater quantities of white lead were used in paint manufacture and many old buildings still have layers of this toxic pigment on their walls and woodwork. The manufacture of white lead in Britain today takes about 3000 tonnes of the metal annually, compared to 10 times as much in the 1930s. However, only a small amount of today's white lead is actually used for paint-making. The bulk goes into plastics.

A lead pigment still used in large quantities is red lead. Along with the more recently developed calcium plumbate, it is used as a rust inhibitor for iron and steel. An idea of the quantity of such lead paints used in protecting steel and iron structures is given by the fact that 91 tonnes are used each year in the continuous painting that is carried out on the Sydney Harbour Bridge in Australia.

A major use of lead salts is in glazing of ceramics. About 3000 tonnes of lead, mostly as lead bisilicate, are used for this purpose in Britain each year. Lead salts are also used extensively in glass manufacture, both for high-quality crystal and for television tubes and fluorescent lights. Though, as has been mentioned, the replacement of lead pipes and cisterns by plastics has led to a reduction of the use of the metal in plumbing, plastics such as polyvinyl chloride (PVC) require lead salts as stabilisers.

There are many other minor uses of lead which between them use about 3000 tonnes of the metal annually in Britain. They include driers in paint and printing inks, agricultural fungicides and insecticides, and antifouling compounds for hulls of ships. Such uses seem insignificant when compared with the major use of lead compounds today, as petroleum additives. In Britain alone, about 40 000 tonnes of metallic lead are converted annually into tetraethyl and tetramethyl lead for this purpose. Total world production in 1973 was 10 times the British level. The consequences of such use for world health, and in particular for increasing the levels of lead in our diet, are considerable and have been the subject of vigorous debate for many years.

LEAD IN THE HUMAN BODY

Lead has been listed by Schroeder and Nason among the 'abnormal trace elements of interest in the human body and blood'[5]. It is certainly of interest, but is hardly abnormal, because lead is present in practically every organ and tissue of the human body. The total amount present varies with age, occupation and even race. It has been estimated that western 'Reference Man', a 70 kg male who has not been exposed to excessive amounts of environmental lead, contains between 100 and 400 mg of lead with an average of 120 mg, or 1.7 µg/g of tissue. Of this, 1.4 mg is in blood and more than 100 mg, or some 92 per cent of the total, in bone.[5] In Britain the concentration of lead in bones of men and women over 16 years of age ranges from 9–34 mg/kg. Liver contains about 1 mg/kg and kidney somewhat less. Similar levels have been observed by other workers in Japan and the USA. In Britain the level ranges between 0.02 and 0.8 mg/kg.[6]

We probably begin life with a small body store of lead. Transfer across the placenta from the mother to the foetus has been shown to occur.

The lead content of the body increases with age, as our exposure to environmental lead is prolonged. This lifelong build-up is mainly in bone, which acts as a reservoir for the metal. It is not totally and permanently fixed there, but under certain conditions of stress may be released into the bloodstream.

After birth we absorb our body's lead either from our food and drink or from the air. Studies by Kehoe and others[7] have indicated that about 10 per cent of ingested lead is absorbed from the gastro-intestinal tract of adults. There are indications that the amount absorbed by children may be higher. As much as 53 per cent of dietary lead was absorbed by a group of children ranging in age from 3 months to 8 years.[8] Newborn animals have also been shown to have a high absorption rate.

Several dietary factors can affect the level of absorption. A low body-calcium status results in an increased absorption of lead. Uptake of both calcium and lead is increased by vitamin D. Iron deficiency may also promote lead absorption, as will fasting. A diet high in carbohydrates but lacking in protein can have a similar effect.

Once absorbed into the bloodstream, lead is transported round the body like most other heavy metals, by being attached to blood cells and constituents of the plasma. It is distributed to form an exchangeable compartment (blood and soft tissue) and a storage compartment (bone). Lead in blood is bound mainly to the erythrocytes, where its concentration is about 16 times higher than in plasma. Some of the lead absorbed is transported to the brain. Metallic lead, however, does not accumulate there to any great extent. Tetraethyl lead, on the other hand, appears to be retained preferentially in brain tissue, where it is partially broken down to form the triethyl derivative.

Lead is eliminated from the body in faeces, as well as in urine and in sweat. About 90 per cent of ingested lead is lost in faeces. Of the 10 per cent that is actually absorbed from the diet, about three-quarters will eventually leave the body in urine. Other excretory pathways are gastro-intestinal secretions, hair, nails and sweat. Schroeder believes that the latter pathway is of considerable importance and says that the 'sense of well-being resulting from sauna baths, vigorous exercise or exposure to the hot and low-lead environment of the tropics in the result of negative lead balances induced by sweating'.

Lead finds its way out of the body in milk. Levels of as high as 0.12 mg/litre have been found in human milk in Japan.[9] Figures published by Murthy and Rhea[10] for the USA are 10 times lower. Increases in the lead content of human as well as cows' milk have been found in areas where environmental lead is abnormally high. While there is no evidence that human infants can be poisoned by lead in mother's milk, poisoning has been produced in suckling rats and mice by exposing their mothers to lead.

Biological half-time is an important concept. When uptake of a metal exceeds elimination rate, it will be accumulated by the tissues. A steady state is reached when uptake and loss balance each other. This steady state may be followed by an elimination stage, and the rate at which such elimination takes place is significant. The biological half-time ($T_{1/2}$) gives a measure of how long it takes to reduce accumulation in an organ or the whole body to half the original level. Different estimates have been made of biological half-times for lead in the human body. The differences are due mainly to our lack of precise knowledge of how lead is stored. Tsuchiya and Sugita,[11] assuming an exponential one-compartment model, obtained an estimate of 5 years for the total body $T_{1/2}$. The Task Group on Metal Accumulation[12] estimated a $T_{1/2}$ of 10 years for lead in human bones. This calculation and a more recent one[13] which gives a somewhat higher $T_{1/2}$ are based on a three-compartment model consisting of blood-lead and some rapidly exchanging soft tissue with a $T_{1/2}$ of about 19 days; soft tissue and a rapidly exchangeable bone fraction with $T_{1/2}$ about 21 days; and the skeleton with $T_{1/2}$ of about 20 years.

There is apparently a similarity between the metabolism of lead and of calcium. Both metals are found in bone crystal structure, which consists mainly of calcium phosphate. It was once believed that with time, lead became buried deeper and deeper within the bone structure and was held there in permanent state. However, there is now evidence that such 'burying' taken place only over a very long time and that a potentially dangerous pool of available lead persists for many months after exposure to high levels of lead. It is also possible that even 'fixed' lead may be mobilised from bone under certain conditions of shock and illness.

Considerable variations have been found between levels of lead in bone of persons of different ages and sexes and from different parts of the world. Lead in the skeletons of 258 Americans varied from 7.5 to 195 mg/kg, with an average of 100 mg/kg.[14] In contrast, skeletons of Peruvian Indians dating from about AD 1200 had less than 5 mg/kg. Third-century Polish skeletons had between 5 and 10 mg/kg in contrast to modern Poles with about 12 mg/kg.

Lead in hair has been used as an indicator of exposure to heavy metal contamination and to assess levels of dietary uptake. However, there is considerable lack of agreement about the reliability of such assessments. Hair of children, for instance, has been used in an attempt to examine the level of exposure to lead in the USA over the past century.[15] Locks of hair preserved in lockets from children who lived between 1871 and 1923, as well as fresh samples of modern hair, were analysed. The 'antique' hair had a mean of 164 mg/kg of lead compared to 16 mg/kg in modern hair. This reduction, it was claimed, resulted from changes in the American way of life. These changes included the fact that drinking-water was no longer collected from leaded roofs and poorly glazed earthenware vessels were no longer used for holding food and beverages. However, others[16] have shown that the assessment of lead contamination from hair levels is not a straightforward matter. Age and sex, as well as the presence of other metals in the diet, can affect accumulation. The simultaneous presence of cadmium, for instance, can enhance the uptake of lead considerably. A survey of hair-lead carried out among students and others in Oxford, England, showed[17] that there was considerable variation in levels even in samples from those who lived in the same area and had, apparently, a similar way of life. Values ranged from 0.2 to 54.2 mg/kg, with a mean of 5.2 mg/kg.

Biological Effects of Lead

The symptoms of acute lead poisoning following the ingestion of large quantities of the metal are well known. Less is known about chronic effects which may occur after the accumulation of lead in the body over a long period of time. The major effects are related to four organ systems: haemopoietic, nervous, gastro-intestinal and renal. Acute lead poisoning usually manifests itself in gastro-intestinal effects. Anorexia, dyspepsia and constipation may be followed by an attack of colic with intense paroxysmal abdominal pain. This is the 'dry gripes' or the 'Devonshire colic' formerly known to cider and spirit makers who used lead-lined vessels in their fermenting and distillation processes. At times the pain from lead colic is so severe that it can be taken for acute appendicitis. Lead encephalopathy in adults is rare but it occurs more frequently in children. It has been observed in children with pica in the USA[18] and in African children living in the vicinity of a zinc–lead smelter.[19] Some children may have anaemia and milk colic prior to the onset of the acute encephalopathy syndrome, which includes vomiting, apathy, drowsiness, stupor, ataxia, hyperactivity and other neurological symptoms.[20]

Today, gross symptoms of lead poisoning are seldom met with except in those exposed to extreme occupational hazards. Attention is largely centred on subclinical levels of poisoning, the chronic condition that may follow on residence in a high-lead environment or ingestion over a long period of time of small amounts of the metal in food. One of the effects of such subclinical poisoning is interference with the pathway of haem biosynthesis. Mild anaemia is often observed among occupationally exposed workers. Though it is seldom noted today, the characteristic 'pallor' of lead workers was first described as long ago as 1831 by the French physician Laennec. This lead-induced anaemia is caused by the combined effect of the inhibition of haemoglobin synthesis and shortened life span of the circulating erythrocytes. It is seldom of any serious consequence with regard to health but is of interest from the diagnostic point of view. Many stages in the pathway of haem synthesis are inhibited by lead. Delta-aminolevulinic acid dehydrase (ALA-D), which catalyses the formation of porphobilinogen from σ-aminolevulinic acid (ALA), and haem synthetase (haem-S), which incorporates iron into protoporphyrin IX (PP IX), are the enzymes most affected. However, other stages can also be blocked.

Measurement of the activity of ALA-D activity, as indicated by concentration of ALA in the blood, provides a sensitive clinical test of lead poisoning. It gives an indication of levels of lead uptake long before any obvious physical symptoms appear. This is of value in detecting incipient lead poisoning.

The shortening of the life span of erythrocytes may also be a cause of anaemia in lead poisoning. The exact mechanism by which the shortening occurs is not fully known. It is possible that lead causes increased mechanical fragility and reduces osmotic resistance in the blood cells.

Effects of lead on the nervous system, both central and peripheral, are well known. Apart from acute encephalopathy, milder symptoms of the effects of lead on the central nervous system occur, which may include mental deterioration and hyperkinetic or aggressive behaviour. However, it is very difficult to establish a direct relationship between elevated levels of blood-lead and neurophysiological effects, especially if they are subclinical. Cerebral injury due to viral or other causes has been known to result in persistent pica, with a consequent increased intake of lead, thus making high blood-lead levels a consequence rather than a cause of mental deterioration.

Peripheral neuropathy was formerly frequently observed among workers exposed to lead and in others whose intake of the metal in food and drink was excessive. Lead palsy resulting in wrist or foot drop, in which there is actual paralysis of the muscles of the hands and feet, is now seldom observed. During prohibition days of the 1930s in the USA several cases of ankle drop were reported among dancers in cabarets where illicit, lead-contaminated 'bath-tub' gin was consumed.

The cause of this palsy has been extensively studied. It is related to muscular fatigue. Muscle and nerve action depend on a balanced flow of calcium across cell membranes. Lead interferes with this movement by forming soluble lead lactate with lactic acid produced during muscle metabolism. This lead lactate readily penetrates into muscle and nerve cells and there it is once more changed—this time by combining with phosphate, a normal constituent of tissues—into an insoluble form. This lead phosphate settles like a barrier on the surfaces of cells and blocks free passage of calcium. The neuromuscular effects of lead-induced palsy are a consequence of this blockage.

There is evidence that chronic kidney disease may result from prolonged exposure to even small regular intakes of lead. A WHO report[21] concluded that prolonged exposure to lead with blood levels above 70 μg/100 ml may result in chronic irreversible nephropathy. A report from Australia[22] seems to support this conclusion. This notes that the death rate from chronic nephritis in Queensland is far higher than elsewhere in the subcontinent. It is suggested that the cause of the high levels of nephritis deaths was childhood lead poisoning due to drinking water collected from roofs which had been painted with lead-based paints. Of 401 cases of childhood lead poisoning, follow-up studies showed that 165 died under the age of 40, and 101 of these deaths were attributed to kidney failure. Nevertheless, there is considerable controversy as to whether in fact childhood exposure to lead can be associated with the occurrence of nephropathy in later life.

Carcinogenic Effects of Lead

Stocks and Davies,[23] following a 10 year study of the incidence of various forms of cancer in north Wales, came to the conclusion that there was a definite link between the amount of metals, including lead, in vegetables, as well as in the soil on which they were grown, and cancer in humans. However, a report published by the International Agency for Research on Cancer in 1972 concluded that there was no evidence to suggest that lead salts cause cancer in man. While tumours have been induced in rats and mice by oral and parenteral administration of various lead compounds, these have only resulted after high doses. The equivalent dosage in man would have to be 550 mg of lead a day.[24]

ALKYL LEAD

A particular problem is presented by tetraethyl lead (TEL) and the other alkyl lead compounds. As has already been mentioned, large quantities of TEL are used in motor fuels. The amount used varies in different countries—in some its use is restricted considerably. However, in general, TEL makes up about 0.1 per cent of the fuel content. It is an acutely toxic compound—far more dangerous than metallic lead or its inorganic compounds—and the poisoning it causes is quite different from that due to inorganic lead. TEL is very volatile and its absorption is usually by inhalation, but the liquid is easily taken in by the skin as well as by the gastro-intestinal tract. After absorption TEL is distributed to various tissues, particularly to the brain and other organs, decomposing into triethyl lead and minute amounts of inorganic lead.

It is unlikely that TEL contamination of food occurs in more than isolated cases, but accidental drinking of contaminated beverages has occurred. Most known cases of TEL poisoning have resulted from inhalation during cleaning of petroleum storage tanks or from the misuse of gasoline as a cleaning fluid in badly ventilated work-rooms. The earliest symptom of alkyl lead intoxication is insomnia and the main organ affected is the nervous system. Poisoning is usually acute, developing into toxic psychosis, and may result in death.

Though metallic lead as well as its organic forms have been found in the brain and other tissues of persons who have died of TEL poisoning, this is not always the case and it would appear that the toxic action is due not to lead itself but to the organic part of the antiknock compounds.[25]

Most of the TEL in motor fuels is decomposed by combustion into inorganic lead compounds such as lead halides and oxides. However, very small amounts of organic lead do escape into the ambient air. A US Department of Health, Education and Welfare survey[26] found that TEL concentrations in the atmosphere of urban areas did not reach 10 per cent of the inorganic lead levels and were too low to have adverse effects on general health. Nevertheless, TEL in motor fuels does contribute a very great deal of inorganic lead to the environment—in the UK as much as 10 000 tonnes a year. There is evidence that very large amounts of lead do occur in soils and herbage near highways, as has been shown in the USA,[27] in Birmingham, England[28] and in some other places. However, the extent of such contamination and its actual significance for the food industry seem to be quite limited. A study carried out in Italy showed that grapes growing immediately beside a busy roadway and on the edge of a parking area did have an increased level of lead compared to normal values, but even a few metres away, the influence of traffic was found to be negligible.[29] While lead levels in grass close to an autobahn were increased by exposure to exhaust fumes, milk of cows which grazed in the contaminated fields had normal levels of the metal.[30]

A striking confirmation of this finding was provided by a study carried out at the Rothamsted Experimental Station, Hertfordshire, England.[31] Samples of topsoil collected regularly at the station since 1881 were analysed for lead. Though the station is not on the immediate verge of a busy highway, it does lie within 2 km of three major trunk roads and is 42 km from the centre of London. Thus its soil and herbage lead levels can be expected to reflect changes in the background level of the metal in this industrialised and heavily populated part of England. The results showed that levels had only increased slightly during this century. Herbage samples which were collected over a similar period of time also failed to show any noticeable increase in lead content. It is particularly significant that the results do not in any way reflect the introduction of organic lead antiknock petrol additives in the 1930s or the increase in motor traffic in later years.

LEAD IN ROAD RUN-OFF

While several other studies support the view that the bulk of lead emitted in automobile exhaust is deposited within a short distance of the point of emission and that such emission does not, directly, lead to a significant increase in background levels of the metal, a report prepared by scientists working for the Oklahoma City Water Quality Control Service in the USA[32] showed that exhaust lead could contaminate the environment indirectly on an unexpectedly large scale. They found that while upwards of 50 per cent of the lead in antiknock additives is deposited on the road bed, most of this can be carried away in the run-off from the streets and find its way via the sewers into fresh-water sources. This contamination could be particularly serious after a prolonged dry period during which lead accumulated on the street and was subsequently carried to surface streams when the drought broke.

Concern has been expressed at the possibility of the contamination of food on display in shops and stores exposed to lead-containing road dusts. However, a report from Germany[33] indicates that such contamination is not significant. Fruit and vegetables displayed in front of a shop in a busy street were examined for lead pick-up. The increase in metal content of the foodstuffs was small and most of this was found to be easily removed by washing. In addition, it was observed that contamination could be prevented by packaging fruit and vegetables in plastic bags or covering them with plastic sheets during display.

LEAD CONTAMINATION OF FOOD AND DRINK
FROM OTHER CAUSES

Lead is a normal ingredient of our diet, whether we like it or not. Different estimates have been made of the amount we ingest daily but actual intake depends to some extent on where we live, the quality and origins of the food and drink we consume and other factors. Kehoe[34] calculated that the average American consumed about 0.3 mg/day of lead in food, with about another 0.1 mg coming from water and other beverages and from atmospheric pollution. This is probably about right for average intake in an industrialised country. Figures for the UK suggest a total intake of lead in food and drink for the average citizen of about 0.2 mg/day.[35] A 1976 WHO Task Group concluded that a fair estimate of daily intake of lead via food was 200–300 μg.[36]

Only about 10 per cent of this ingested lead will actually be absorbed from the alimentary canal. Thus, though there will be differences between absorption levels in different populations and individuals, on average some 20–30 μg of lead will enter the bloodstream from food each day.

In the case of persons with no known high exposure, blood-lead levels, resulting mainly from ingested food, are fairly uniform throughout the world, ranging from between 10 and 35 μg/100 ml on average.[37] These levels are in agreement with the WHO estimates of the relationship of blood-lead to dietary intake of 5.4–18.3 for men and 4.4–13 μg/100 μg oral lead intake per day for women.

Clarkson[38] believes that a level of blood-lead between 50 and 80 μg/100 ml constitutes the top of the 'no-effect' level, at which the metal seems to produce no significant clinical effect in adult man. To reach such levels, a man would have to ingest 8 mg of lead each day. Allowing a safety factor of 10, the maximum permissible daily total intake according to Clarkson should be 0.8 mg. This is considered a generous allowance by many.

Table 8 shows the average lead content of a selection of foods. Similar figures have been reported in many countries. It is evident that certain foods can contribute more than others to the total intake, and official monitoring keeps a close watch in order to forestall any excessive intake due to changes in dietary habits or other factors. In the UK a wide-ranging programme of sampling and monitoring for levels of lead and other heavy metals in food is carried out. Results are published regularly and on the basis of such findings permitted levels of heavy metals in food are established. Where necessary these allowances are modified as new observations either indicate a need for a reduction of intake of certain high-lead foods or that earlier norms in certain cases were too restrictive and in practice can be

TABLE 8
LEAD CONTENT OF SOME FOODS[35, 39]

Food	Mean (mg/kg)	Range (mg/kg)
Cereals	0.17	< 0.01–0.81
Meat and fish	0.17	< 0.01–0.70
Fruit (fresh)	0.12	< 0.01–0.76
Fruit (canned)	0.40	0.04–10.0
Vegetables (fresh)	0.22	< 0.01–1.5
Vegetables (canned)	0.24	0.01–1.5
Milk	0.03	< 0.01–0.08

relaxed. The latter is by no means common, though it has occurred in the case of lead in game. The most recent recommendation of the Committee on Heavy Metal Contamination of Food in the UK has in fact been for a general reduction by 50 per cent of the levels of lead permitted in most foods.

General monitoring does not, unfortunately, give the whole picture of trace metal contamination of food. Certain important features appear when we look in detail at individual foodstuffs and consider particular problem areas.

Lead in Meat

Meat, fish and other animal products have been implicated in a number of serious cases of poisoning due to the transmission of metals from the environment to man. Fish and meat have been found to be dangerously contaminated with mercury, and high levels of cadmium have also been found in free-swimming fish as well as in shellfish. However, though lead is accumulated by oysters and crustaceans in polluted waters, there is no strong evidence that fish or indeed any other animal has been the vehicle by which lead poisoning is transmitted to man.

Lead in Game Meat

The recent recommendation by the UK Joint Committee on Heavy Metal Contamination of Food, that the permitted level of lead in game and game pâté be increased from 5 to 10 mg/kg 'not including discrete lead shot', draws attention to the fact that there is one type of meat which is readily available on the market in some countries which is capable of causing lead poisoning if eaten in sufficient quantity. Game, the rabbits and hares, the ducks, geese, partridges and pheasants which for most of us are infrequent special items in the diet, are more often than not killed by lead pellets shot from a gun. The shot can lodge in the flesh and if not noticed during preparation of the meal may be consumed. But apart from this, sometimes, in the case of birds especially, the animal may have been already carrying a high load of lead even before it was brought down by the hunter. A staggering 15 billion lead shot are fired each hunting season over the duck and geese hunting grounds of North America alone.[40] Only about one-fifth of these particles actually strike a bird. The remainder fall into the water or on to vegetation or the ground and there may be consumed by feeding birds and other animals. Lead-poisoned birds are frequently recovered by Wildlife and Fisheries agents of the USA and Canada. No doubt quite a proportion of the birds killed by hunters and eaten by them, or by others who subsequently purchased them, contain high levels of lead. There is less information about the situation in the UK and other parts of the world, but there is sufficient evidence to show that game birds and animals can be a significant source of lead intake in these countries as well as in America.

It is because of the impossibility of restricting this source of contamination without banning the whole trade in game, or at least insisting that animals be killed by some other means than by gunshot, that the UK regulations have been relaxed. It is clearly accepted that very few people will eat game regularly or in such large amounts that they will suffer lead poisoning as a result. However, it would be unwise to assume that in the case of infants and young children, game does not represent a better avoided hazard for health.

Lead in Beverages

Most natural waters contain about 5 μg of lead per litre. The WHO has recommended a maximum permissible limit of 50 μg/litre for drinking-water. This level has been adopted in Britain, the USA and some other countries. Most municipal supplies will contain well below this limit but, as recent evidence shows, this is not always the case. A survey carried out in Liverpool found that of 47 samples from public water supplies, 22 had concentrations at or above the WHO recommended level. Results from similar studies in rural Scotland as well as from the city of Glasgow indicate that high lead levels in water supplies are not uncommon.[41] Such contamination is not confined to the UK, as results of investigations of municipal supplies in Boston and Seattle in the USA show.[42]

Lead contamination of water supplies may be due to pollution of rivers, wells and other sources by industrial and municipal waste output. However, it is most often caused by the use of lead in plumbing systems. Lead plumbing was almost universal some decades ago but has now largely been replaced by other metals and by plastic. Older properties still frequently have such systems even though, in the UK at least, local authorities are replacing such systems wherever possible.

The use of lead plumbing systems is an especially serious problem where the water supply is 'soft' with a low pH. Such water is plumbo-solvent and can dissolve large amounts of lead from the system. Hard water, on the other hand, of high pH and containing dissolved salts of calcium and magnesium, forms 'scale' within the system which prevents solution of lead and other metals. Acid moorland water, such as is common in rural Scotland and Wales, may dissolve as much as 25 mg/litre of lead—several thousand times the WHO recommended limit. The danger is increased when water has been standing in contact with metal for several hours and not simply running rapidly through lead pipes. The recent UK Department of the Environment survey referred to above found that 9 per cent of all homes tested had more than 100 μg/litre of lead and another 20 per cent more than 50 μg/litre (the WHO maximum permitted level) in 'first-run' tap-water. These percentages were reduced to 4 and 10 per cent in 'daytime' running water. In the light of such differences between 'first-run' and 'daytime' tap-water it is good policy, especially where the feeding of young children is concerned, to allow water to run out of the tap for some minutes before using it to prepare food or beverages. The lead-enriched overnight water will flow away down the drain and fresh, less polluted water will reach the tap. This precaution is not a trivial matter, as H. Egan of the Laboratory of the Government Chemist of the UK noted.[43] He has calculated that 'an untreated private supply with the maximum amount of lead allowed would, on the assumption that 10 per cent of lead is absorbed and that two early-morning cups of tea are consumed, give a daily uptake of about 20 μg'.

In the modern home, lead pipes and tanks are seldom used and have been replaced by plastics. Though lead salts are used as stabilisers in the making of polyvinyl chloride, there is no evidence that the metal can be leached out in dangerous amounts in normal use. However, in newly installed PVC systems, lead may appear when water first runs through. It is advisable for safety that plastic plumbing systems should be thoroughly flushed out over a period of several hours before the water is used for human consumption.

Lead in Alcoholic Beverages
The use of lead pipes and other equipment in breweries and cider factories in the past was often responsible for contamination of beer and cider. Today's plants use stainless steel or other lead-free material, and significant levels of lead contamination are seldom encountered in commercially prepared alcoholic beverages. A study of a range of UK and imported bottled and canned lagers[44] found less than 0.06 mg/litre of lead in 144 samples examined. Similar low levels of lead have been found in other canned beverages. The cans used are usually made of extruded aluminium. Some, however, may have been sealed with solder, but modern production techniques are such that lead contamination of the can contents seldom occurs. Contamination of wines with lead, especially if they are prepared in traditional ways, is not uncommon.

The use of metal foil caps has been shown to cause some contamination at times. The use of lead and lead alloys in construction and repairing of fermentation and other equipment has also been known to result in contamination. The use of bronze taps on casks and tanks has been found to cause lead contamination of wine.

Contamination of potable spirits on a smaller scale due to the use of tinned measures has been reported on a number of occasions. One such measure tested by the author was found to release 5 mg/litre of lead when filled with a dilute acetic acid solution for less than one hour. Though such contamination is uncommon today, it is interesting to note how recent is this change for the better. In 1954 the British Food Standards Committee reported that lead was one of the most widespread and serious of metallic contaminants of drink. This fact the Committee attributed largely to the use of lead and lead alloys in the equipment used in the industry as well as to the use of lead insecticides in agriculture. In 1961 a general limit of 0.2 mg/litre was set for ready-to-drink non-alcoholic beverages and as a special concession until 1964, 1.0 mg/litre for beer, cider and perry. The present regulations have reduced this permitted level. The Committee also recommended that the use of lead piping and lead-containing equipment should be reduced. The result of such recommendations in the UK is that today commercially produced beverages are normally free of more than trace amounts of lead. Since, however, lead arsenate is still used as an insecticide in apple and pear orchards, unless particular precautions are taken, ciders and perries can still be lead-contaminated. It is interesting to note that as long ago as 1767 Sir George Baker, in *An Essay on the Endemic Colic of Devonshire*, blamed the use of lead-lined troughs in the manufacture of cider for this illness.

Lead contamination of beverages still occurs frequently today in home production. Because of the use of readily available and non-food-quality equipment, illicit spirits are often found to contain large amounts of lead and other toxic metals. Poisoning from the consumption of such beverages is not uncommon.[45] The presence of lead in a product is usually related to the use of lead pipes or solder in the apparatus. It also may result from storage of the product in unsuitable metal or glazed earthenware containers. Fatalities due to this latter cause have been reported in recent years. Apple juice stored in an earthenware jar caused the death by lead poisoning of a young boy in Canada.[46] After three days' storage in the hand-crafted jug the juice was found to contain 1300 mg/litre of lead. This is several thousand times the permitted level in any country and it caused the death of the boy within three days of admission to hospital.

As a result of this case an investigation was subsequently carried out on earthenware pottery available in Canadian stores as well as on others produced under home-craft and art-class conditions. More than half the several hundred samples examined were found to be unfit for culinary use because of the amount of lead that could be leached from them by mildly acid conditions. Between 10 and 25 per cent would have been capable of causing severe poisoning such as that from which the boy died. These results are by no means unique, as investigations carried out elsewhere have shown. In 1960 a case of lead poisoning due to drinking home-made wine which had been stored in an earthenware container was reported in Britain.[47] Forty persons were poisoned in a similar manner in Yugoslavia.[48] An American medical doctor suffered lead poisoning from the constant use of a ceramic mug made by his son in a craft class.[49] Many other similar instances could be given, all of which would serve to confirm the wise advice that, unless pottery and other ceramic containers have been subjected to a standard leaching test and shown not to cause contamination of their contents, they should be relegated to the function of ornaments and souvenirs and not used for culinary and other food uses.

Contamination of foods and beverages due to the use of unsuitable materials for cooking and handling equipment has already been discussed in an earlier chapter. Tin, which contains lead, and is used to surface copper and iron utensils, as well as ceramic and other glazes on cast-iron casseroles, may, as we have seen, cause contamination under certain conditions.

ANALYSIS OF FOODSTUFFS FOR LEAD

Both wet digestion and dry ashing can be used for sample pre-treatment for the estimation of lead in food. Care must be taken in the wet process as the use of sulphuric acid can result in low recoveries, due to the formation of insoluble sulphate. Recovery in dry ashing is good, provided the temperature is maintained at around 500 °C.

Dithiozone (diphenylthiocarbazone) forms a red complex with lead. This can be used for a spectrophotometric determination of the metal. A recovery of 90 per cent with a limit of detection of 0.1 mg for lead in canned food has been reported for this method.[50]

Atomic absorption spectrophotometry is the current method of choice for lead in foods. Because of the widespread distribution of lead, care must be taken to clean all glassware with nitric acid and to use only lead-free reagents. Matrix effects occur but can be overcome by preliminary separation of the lead from the digest or ash. This can be done by extraction into isobutyl methylketone (IBMK) using ammonium pyrrolidine dithiocarbonate (APDC) as a chelating agent. The use of background correction also improves the technique. Other metal ions do not interfere with the estimation in an air–acetylene flame. High concentrations of anions, however, such as phosphate, acetate and carbonate, suppress lead absorbance significantly.

Considerable work has been done on the estimation of lead in blood. Numerous inter-laboratory comparisons have been reported.[51] Reference has already been made to some of the micromethods employed, such as use of the Delves cup. In spite of these efforts a surprising lack of agreement and a high level of inaccuracy have been reported, especially in the analysis of small-size (microlitre) samples. Lead analysis in larger food samples appears to be more reliable.

Electro-analytical methods using anodic stripping and polarography are also used for determining lead in food. Neutron activation analysis, while desirable because of its non-destructive nature, is not used extensively as yet because of the need for a fast neutron source.[52]

6

Mercury and Cadmium

MERCURY

Mercury is one of the ancient metals. It is fairly easily extracted from its ores and, in spite of its limited practical use, has been sought and valued for many hundreds of years. The Spanish conquistadors brought it back to Europe from the New World to augment supplies from their own rich mine, after which they named their Mexican source New Almaden. But it is a peculiar metal, useless for making weapons or tools, for it is a liquid under normal conditions. It was because of this strange property that it was called quicksilver. For the Renaissance world its value lay largely in its medicinal properties—some real, many imagined—in its use as an amalgam with some other metals and as a means of 'silvering' mirrors. For the medieval alchemist mercury had its own peculiar value and it played an important part in the search for the 'philosopher's stone', the mysterious substance which would allow the transmutation of base metals into gold. It was the modern alchemist, the technologist of today, who showed that mercury was indeed capable of taking part in chemical transmutations, as a catalyst in a variety of industrial and laboratory reactions, some of great economic importance. It is to a large extent as a result of its catalytic and chemical properties that mercury finds use today, though it is also of importance, because of its physical property of conductivity, in the electrical industry.

From early times men knew that this attractive and interesting liquid metal had sinister side-effects. Those who worked with mercury sometimes suffered from distinctive illnesses. Georgius Agricola tells of the 'quicksilver disease' of the miners in the Hartz Mountains, and miners and refiners elsewhere were also known to suffer from similar complaints. The saying 'as mad as a hatter' recalls the terrible fate of those workers who used mercury to prepare the felt for making beaver hats in former days. Prolonged exposure to mercury vapour in poorly ventilated workshops resulted in progressive mental deterioration.

Though hazards of this kind have been eliminated almost everywhere by modern industrial hygiene practices, we today have once again been made aware of the dangers to health which may result from the use of this metal. Compounds of mercury consumed in fish, cereals and other foodstuffs have resulted in numerous fatalities and thousands of cases of painful and debilitating illness in recent years. Many would hold that mercury is among the most dangerous of all the metal contaminants we are likely to meet in our food. Worldwide attention among food and toxicology experts is focused on the metal, and regular monitoring of levels in the human diet is being carried out in every developed country.

CHEMICAL AND PHYSICAL PROPERTIES

Mercury, given the chemical symbol Hg after its Greek name *hydrargyrum* (quicksilver) is element number 80 in the Periodic Table. Its atomic weight is 200.6 and its density 13.6, making it one of the heavier of the metals. It is liquid over a wide range of temperatures, from its melting-point of $-38.9°C$ to its boiling-point of $356.6°C$. Its oxidation states are 1 and 2.

Elemental mercury is rather volatile and a saturated atmosphere of the vapour contains approximately $18 mg/m^3$ at $24°C$. It is slightly soluble in water, and also in lipids where solubility is of the order of $5–50 mg/litre$. The metal oxidises rapidly in the presence of air. It can also form sulphates, halides and nitrates which are soluble in water. In aqueous solution an equilibrium is formed between the Hg^0, Hg_2^{2+} (mercurous) and Hg^{2+} (mercuric) states. The proportions of the different oxidation states are determined by the redox potential of the solution and the presence of complexing substances. Hg^{2+} ions are able to form many stable complexes with biological compounds, especially through $-SH$ (sulphydryl) groups. In aqueous solution, four different combinations with chlorine are formed: $HgCl^+$, $HgCl_2$, $HgCl_3^-$ and $HgCl_4^{2-}$. Mercurous mercury is rather unstable and in the presence of biological material tends to dissociate to one atom of metallic mercury and an Hg^{2+} ion.

ORGANIC COMPOUNDS OF MERCURY

Volatile compounds are formed between alkyl mercuric compounds and the halogens. These are highly toxic. Less volatile are the hydroxide and nitrate of the short-chain alkyl mercuric compounds. Methyl and ethyl mercury chloride have a high solubility in solvents and lipids. The methyl mercury group has a high affinity for sulphydryl groups and it is through them that methyl mercury binds to protein in living organisms.

ENVIRONMENTAL DISTRIBUTION

Mercury is not a common element in the environment. It is only the sixty-second most abundant element in the earth's crust, with an average concentration of about 0.5 mg/kg. Its distribution is not homogeneous, but occurs in large quantities only in certain areas. The world's principal mercury mines occur in what is known as the 'mercuriferous belt' stretching along the Mid-Atlantic Ridge, the Mediterranean, South and East Asia, and the coastline around the Pacific including New Zealand and Central America.[1]

Famous mines of Europe and the Western world are Idria in Yugoslavia, Almaden in Spain and the New Almaden in California. There are also large mines in the USSR. The metal is produced on a smaller scale in a number of other countries, including Ireland. Apart from its extraction by man in mining operations, large quantities of metallic mercury are released into the environment naturally by volatilisation during earthquakes and volcanic activity. Combustion of fossil fuels also contributes in large measure to the environmental burden of the metal.

Mercury occurs in the earth's crust mainly in the form of various sulphides. Cinnabar, the red sulphide, is the principal ore mined. The ores, which may contain upwards of 70 per cent of the metal, are relatively easily extracted to give the pure metal. The sulphides are roasted in air and the mercury volatilises. The vapour is condensed and the metal, which is a liquid at normal temperature, is collected in flasks.

WORLD PRODUCTION AND USES

Production on a world-wide scale is about 10 000 tonnes per annum. Of this some 25 per cent is used as electrodes in the chlor-alkali industry, 20 per cent in electrical equipment, 15 per cent in paints especially for marine use, 10 per cent in the manufacture of various mercury-containing instruments such as thermometers, 5 per cent in seed dressings and similar agrochemicals, and 3 per cent as mercury amalgam for dentistry. Another quarter of the world production is used in a variety of industrial and military applications such as in detonators, catalysts (for example, in acetaldehyde and polyvinyl chloride manufacture), in the paper pulp industry, pharmaceuticals and cosmetics. Because of their efficient toxic properties, compounds of mercury have been widely used in germicides and pesticides added to paints, plastics and other products. Formerly, extensive use was made of organic mercury compounds for the same purpose and also as antifungal seed dressings, but since the dangers connected with such compounds have been recognised, many countries have banned their use.

In addition to the 10 000 tonnes of mercury brought into the world of technology as a result of mining activities, it has been calculated that a further 10 000 tonnes are released into the environment by the combustion of coal, oil and gas, waste disposal and other industrial activities.[2] Nature herself releases between 30 000 and 150 000 tonnes per year of mercury by degassing from the earth's crust and the oceans.

CHEMICAL CHANGES ON EXPOSURE TO THE ENVIRONMENT

Metallic mercury undergoes a number of chemical changes when it is released into the environment. In soil, sulphur-reducing bacteria may convert mercury into its sulphide. The same transformation can occur under anaerobic conditions in water, but generally in an aquatic system aerobic methylation is the more important transformation. Methyl mercury, CH_3Hg^+, is formed by aquatic micro-organisms both from elemental and from mercuric mercury. From the point of view of the micro-organism, the methylation, which is linked to methionine synthesis,[3] is an efficient way of detoxifying Hg^{2+}. The methylation reaction is likely to occur in upper sedimentary layers on sea or lake bottoms. The methyl mercury formed by micro-organisms, as well as that which is sometimes emitted as a waste product from industry, is rapidly taken up by living organisms and enters the food-chain via plankton filter-feeding bottom invertebrates. A further distribution of mercury from its original location in water may result from this uptake of methyl mercury by living organisms and subsequent degradation. Volatile dimethyl mercury, $(CH_3)_2Hg$, is formed and released into the air. This may be decomposed in the atmosphere by acidic rainwater to monomethyl mercury and thus return to an aquatic environment.[4] It has been shown by the Swedish Expert Group[5] that the local loading of methyl mercury may be considerably increased as a result of such biological and chemical activities following release of mercury from paper pulp and chlor-alkali plants.

MERCURY IN OUR FOOD

Mercury may be present in food in three different forms: as elemental mercury, mercuric mercury and alkyl mercury. The form influences absorption, distribution and biological half-life. In general, however, it may be said that, except in special situations and normally as the result of extreme pollution or misuse of certain compounds, the level of mercury in our diet is very low. There is little danger for most people of mercury poisoning from food. The level of total mercury in animal products, excluding fish, varies from a few micrograms to 50 μg/kg.[6] Fish can exceed these levels considerably.

Metallic mercury is an uncommon contaminant of food. A few known cases have been due to deliberate attempts to poison consumers, as occurred when Israeli oranges were reported to have been injected with metallic mercury by Palestinian activists in 1978. Accidental spillage of mercury from a broken thermometer used in cooking equipment has also been known to occur. There is little danger of toxicity from such ingestion. The metal is poorly absorbed and release of the much more readily absorbed vapour from the surface of the food is unlikely to occur. The metal is quickly eliminated from the body and is of little toxicological significance. Indeed, ingestion of liquid mercury has in the past been recommended medicinally for certain ailments. It has also been ingested in quantity in stage magician's acts without apparent ill effects. Mercury vapour, however, which is readily absorbed from the lungs because of its lipid solubility, is a much more serious problem and is capable of causing both acute and chronic poisoning. However, it is not a problem normally encountered with food and will not be further considered here.

Mercuric mercury ingestion in food is normally due to accident or sometimes to deliberate act. In an earlier age corrosive sublimate (mercuric chloride) was not infrequently used as a poison by murderers and in suicides. Accidental presence in foods would be very unusual and would probably be quickly revealed by its corrosive action. When ingested, less than 10 per cent of the mercuric chloride is absorbed, but a massive dose might cause intestinal damage and result in increased absorption. The greater part of absorbed Hg^{2+} is concentrated in the organs, especially the kidneys. Much of the mercury is excreted by the kidneys and the remainder by the faecal route. Elimination seems to be multiphasic, with an initial rapid half-time, followed by two others of longer duration.

Target organs in acute poisoning are the kidneys and the gastro-intestinal tract. Damage is caused to the mucous membranes of the tract, and severe pain and vomiting occur, which can lead to collapse and death. If the patient survives the initial gastro-intestinal damage, kidney failure can result within 24 hours due to necrosis of the proximal tubular epithelium.

Chronic poisoning is unlikely with Hg^{2+} except if mercury vapour is also present. In such cases renal damage occurs and, in addition, inflammatory changes and black lines have been observed on the gums. The lethal dose of Hg^{2+} is probably of the order of 1 g. Kidney damage has been observed when levels of 10–70 mg/kg have been found in the organ. Normal levels in kidneys, in the absence of mercury poisoning, are between 0.1 and 3mg/kg.[7]

Organic Mercury Compounds

Alkyl mercury presents a far more serious problem than either metallic or mercuric mercury as a contaminant of food. As has been noted already, alkyl mercury formed in marine and freshwater sediments enters the food-chain and may be concentrated via filter-feeding organisms and through them in fish, either by eating the plankton or directly through the gills. The concentration factor in fish (such as pike) may exceed 3000, compared to the concentration of mercury in the surrounding water. Mercury is abundant in many polluted waters, and even in the absence of industrial outflow surface water may contain as much as 200 ng/litre, compared to ocean water with 30 ng/litre.[8] It has been estimated that the atmosphere keeps topping up the mercury levels in water, through precipitation, by approximately 50 000 tonnes a year. Thus it is clear that there is a plentiful supply of mercury available in water for conversion into alkyl compounds and concentration in fish.

Accumulation in fish is related to age and size. Large tuna over 60 kg in weight may have levels of organic mercury of up to 1 mg/kg in muscle. This compares with terrestrial animals whose muscles rarely have more than 50 μg/kg and generally average 20 μg/kg. Fish living in highly polluted water may have more than 10 mg/kg. In British coastal waters average levels of 210 μg/kg have been found for England and Wales and 70 μg/kg for Scotland. However, in the most polluted waters, such as at the mouth of the Thames and on the Lancashire coast, samples contained as much as 510 and 640 μg/kg.[9] These levels are well below contamination levels found in some Swedish lakes and in parts of Japan where methyl mercury poisonings have occurred.

The use of alkyl mercury compounds in agriculture sometimes results in animals ingesting high levels of such compounds, with consequent contamination of human food. Most mercury consumed by man, at least in animal products, is in the form of methyl mercury. The bulk of this actually occurs in fish and so the level of fish consumption by man will be a measure of his intake of organic mercury. In communities where fish consumption is high, especially if these fish are taken from polluted water, the intake of alkyl mercury can reach serious levels. This occurred at Minamata Bay in Japan in the mid 1950s. Pollution of the fishing grounds by industrial methyl mercury emission led to a large-scale epidemic, the first to be investigated thoroughly and to have attracted world-wide concern. Fish and shellfish were found to have levels of methyl mercury up to 29 mg/kg. The result was that daily intake by the local fishing population was as much as 30 μg or higher, compared to the intake from average diets, which is about 10 μg/day. The tragedy of Minamata is that, in spite of indications of what was happening and warnings of the consequences if steps were not taken to prevent further contamination by methyl mercury (a waste product from an acetaldehyde plant which was using an inorganic mercury catalyst), pollution continued until into the 1970s. By February 1971 the total number of poisoning cases was 121, with 46 fatalities. Twenty-two of the cases were congenital. These were infants with cerebral involvement (palsy and retardation) born to mothers who had ingested the contaminated fish. A similar outbreak elsewhere in Japan, at the Agano River, Niigata Prefecture, resulted in 49 cases of poisoning with 6 deaths.

The Japanese tragedy caused considerable concern in other fish-eating countries, especially where mercury pollution of waters occurred. In Sweden the average daily intake of fish was estimated at 30 g. Under normal conditions in the absence of pollution, this level of consumption would result in daily methyl mercury intake of between 1 and 20 μg. However, fish consumption of 200 to 500 g/day is not unknown in some communities. Where the fish were caught in contaminated waters, this high intake could result in an uptake of about 5 mg of methyl mercury each day. Several lakes in Sweden, the USA and Canada, into whose water industrial effluent was released, were found to be heavily polluted and their fish had levels of more than 10 mg/kg of methyl mercury. Fishing was banned in some of them and further emission of polluted effluent reduced.

It is not just in polluted waters that fish can carry high levels of mercury compounds. As has already been noted, some fish have the ability to accumulate methyl mercury even from low levels in water. These include deep-sea species such as tuna and swordfish. It is noteworthy that examination of museum specimens has shown that levels of mercury in tuna and swordfish caught 100 years ago are in the same range as those living today, at about 0.5 mg/kg in tuna but somewhat higher in swordfish. It would appear that the geological cycle rather than industrial emission is responsible for the levels of mercury in these fish.[10] High consumption of such deep-sea fish by the American Samoan population was found by Clarkson et al.[11] to result in a daily intake of 200–300 µg of methyl mercury.

Methyl Mercury in Other Foodstuffs
In a number of exceptional cases, organic mercury compounds intended for agricultural use only have found their way into human foods, with tragic results. In Iraq in 1960 seed wheat which had been treated with methyl mercury was used to make bread. In spite of soaking the grain in water, in a mistaken belief that the procedure removed all danger to health along with the water-soluble red dye which had been used to mark the treated seed, thousands of people who consumed the bread were poisoned. Many died and others were permanently incapacitated. Similar incidents of large-scale alkyl mercury poisonings occurred in Pakistan and in Guatemala a few years after the Iraq case. Many countries now ban the use of such compounds and their production has decreased. Even the use of treated cereals as animal feed caused human poisoning when the animals were later consumed as meat. Apart from such unusual incidents and the few cases of game birds which have fed on methyl mercury-treated seeds, accidental contamination of food by organic mercury compounds is rare. As has been pointed out in an official UK report,[12] most foods, apart from some fish from polluted waters, contain less than 0.005 mg/kg of mercury and thus contribute little to the dietary uptake of mercury.

METABOLISM AND TOXIC EFFECTS OF ORGANIC MERCURY COMPOUNDS

The naturally occurring methyl mercury compounds have been the most studied of all the short-chain organic compounds and about them we have most knowledge. The properties and behaviour of the ethyl mercury compounds are apparently very similar to those of the methyl derivatives.

Methyl mercury ingested in food has been shown to be efficiently absorbed in the intestine. Methyl mercury enters the bloodstream where it is bound to plasma protein. Most of the methyl mercury is accumulated in the erythrocytes. The compound is slowly distributed by the blood to other tissues. The brain shows a special affinity for methyl mercury, and there it can accumulate to about six times the level in other tissues. More than 95 per cent of the mercury in brain tissue has been shown to be in organic form. In other tissues, the organic compound undergoes demethylation to inorganic mercury.[13] Accumulation in the foetus is similar to that in the mother, though foetal brain levels of mercury may be higher.[14] Prenatal poisoning has been shown to occur even when the mothers of affected children showed no clinical signs of intoxication.

Methyl mercury is excreted from the body partly via the kidneys, and the bulk via the liver into bile and thus into faeces. The biological half-time seems to be about 70 days, but evidence from the Iraq catastrophe suggests that the capacity for elimination varies between individuals. A biological half-time of upwards of 190 days was observed in 10 per cent of the population studied.[15]

Much of the methyl mercury in bile is reabsorbed by the gut, but inorganic mercury, which is formed by demethylation in the liver, is largely excreted in faeces.

Methyl mercury is also excreted in breast milk. The concentration in milk is about 5 per cent that in the maternal blood.[16]

There seem to be sharp differences between acute and chronic poisoning due to methyl mercury. Once a toxic dose has been absorbed it will be retained for a long time, causing functional disturbance and damage. There appears to be a latency period between absorption of the dose and appearance of symptoms of poisoning.

Both prenatal and postnatal methyl mercury poisoning are recognised. In prenatal poisoning, an unspecific infantile cerebral palsy involving ataxic motor disturbances and mental symptoms occurs. The clinical signs of postnatal intoxication are characterised initially by sensory disturbances in the limbs, the tongue and around the lips. With increasing intoxication, symptoms become more severe. The central nervous system is damaged irreversibly, resulting in ataxia, tremor, slurred speech, tunnel vision, blindness, loss of hearing and death.[16]

Levels of mercury concentration in the blood give a reliable index of the methyl mercury body burden. Methyl mercury concentration in hair can also be used to measure levels in blood as well as total body burden. Methyl mercury is deposited at the hair pile in proportion to the concentration in the blood at the time.

Thus the mercury levels in different parts of the hair can provide a calendar of blood-mercury uptake over a period of time. In using such results to calculate methyl mercury body burdens under different conditions of dietary intake, allowance must be made for the rate of growth of the hair pile (about 1 cm a month) and the time lag between hair formation and extrusion. The possibility of external contamination of the hair must also be borne in mind.

Attempts have been made during studies of the various methyl mercury poisonings in Japan and Iraq to arrive at a dose–response curve or threshold limit for methyl mercury. These have not been very successful, mainly because of differences in appearance of symptoms in individual patients. However, while allowing for such differences and for the possibility that some populations are more sensitive than others, the WHO has accepted a figure of 5 μg/kg/day as a minimum toxic dose for the compound.[2]

The lowest levels of mercury found in hair in methyl mercury-poisoned persons in Japan was 50 mg/kg. A similar study in Iraq gave the lowest hair values associated with symptoms of methyl mercury poisonings as 120 mg/kg.

ANALYSIS OF FOODSTUFFS FOR MERCURY

The determination of mercury in food and other biological materials requires care and considerable skill. The three principal analytical methods used today are colorimetric; flameless atomic absorption and neutron activation analysis.

The Colorimetric Method
This method is based on the conversion of the metal in the sample to a dithiozone complex, which is extracted by organic solvents and determined colorimetrically. It is a time-consuming operation with a detection limit of about 0.05 mg/kg. It also requires a fairly large (5 g) sample. It has been largely replaced by instrumental methods.

Flameless Atomic Absorption
This method is widely used today for mercury determination, and manufacturers of analytical equipment have put on the market a number of kits which allow standard atomic absorption spectrophotometers to be adapted for cold vapour techniques. Both circulating and non-circulating methods are used. In the former, the mercury content of the sample is determined by measuring the transient absorbance produced when the mercury vapour is led through an absorption cell. This replaces the usual burner in the light path. The circulating methods allow the progressive build-up of mercury vapour until a constant absorbance is attained.[17]

Stannous chloride is used to convert the mercury to its elemental form. The technique is applicable to solutions containing mercury in a form easily reducible by stannous chloride (usually Hg^{2+} or Hg_2^{2+}). The detection limit for the method is 1–5 ng.

Unfortunately, reliability of results depends on care in handling, storage and other preliminary steps. Serious errors result from mishandling, as mercury is volatile and easily lost. Mercury vapour can diffuse through plastics; it can also be absorbed on to surfaces and into materials such as polyethylene, silicone and rubber. In biological materials, bacterial activity can reduce mercuric mercury, resulting in loss of mercury vapour. In aqueous solutions, when the mercury is at low concentrations, it may adhere to the surface of vessels.

Before instrumental analysis of mercury in biological materials, oxidative digestion or combustion in oxygen or some type of extraction may be required. The exact method used will depend on the material being analysed. The determination of mercury in canned fish, for instance, may be performed as follows: approximately 0.5 g samples of homogenised fish are carefully heated in a conical flask with 5 cm³ of concentrated sulphuric acid for one hour on a water bath at 70°C. At the end of the digestion the flask is cooled in ice and 50 cm³ of 6 per cent w/v potassium permanganate solution are carefully added. The flask is reheated for a further two hours at 70°C. After cooling, 15 cm³ of 20 per cent w/v hydroxylammonium chloride solution is added to reduce the excess potassium permanganate. The solution is now ready for analysis.[18]

Alkyl mercury compounds can be identified by thin-layer chromatography or gas–liquid chromatography. The glc technique involves preliminary extraction of alkyl mercury with benzene.[19] Recovery has to be checked for each type of biological material investigated. It is generally above 90 per cent but varies with the sample material. Liver and kidney are more difficult to extract than fish.

Other instrumental analytical methods for mercury are also used. Neutron activation analysis, for example, has a high specificity and accuracy. It has been used effectively on small samples in total diet studies.[20]

CADMIUM

Though cadmium has come into widespread use only relatively recently, it is quite likely that for many centuries this highly toxic metal has been causing food poisoning. Cadmium is usually found associated with other metals in ores, and contamination of zinc and other coating metals with trace amounts of cadmium is not uncommon. Even minute amounts of cadmium are sufficient to cause poisoning and, since the metal is soluble in organic acids, it easily enters foodstuffs. The extreme danger resulting from ingestion of cadmium in food was highlighted by the *itai-itai* disease outbreak in Japan in the late 1960s. This tragedy, due once again, as in the case of Minamata disease, to industrial pollution, drew world-wide attention to cadmium pollution and started a flurry of investigation which is still continuing in many industrialised countries today. Cadmium is among the most serious of all the metal contaminants of food and drink, not only because of its high toxicity but also because of its wide distribution and its numerous and industrially highly valued uses in modern technology.

CHEMICAL AND PHYSICAL PROPERTIES

Cadmium is number 48 in the Periodic Table, with an atomic weight of 112.4. It is a fairly dense (sp. gr. 8.6), silvery white, malleable metal which melts at 320.9 °C and boils at 765 °C. It has an oxidation state of 2. Cadmium forms a number of inorganic compounds, several of which are quite soluble in water, such as the chloride, sulphate and acetate. Cadmium sulphide is only very slightly soluble in water while the oxide is insoluble. A number of organocadmium compounds have been synthesised, but these are very unstable and none has been found to occur in nature. The element forms complexes with organic compounds such as dithiozone and thiocarbamate as well as with proteins through–SH groups.

PRODUCTION

There are no ores of commercial significance which contain cadmium alone. The metal is always obtained as a by-product in the refining of other metals, especially zinc, but also copper and lead. Thus its production is linked entirely to that of these base metals. World production is about 1600 tonnes. This is almost all primary production, since recovery of secondary cadmium from scrap has not yet begun to be of commercial significance.

USES

Cadmium has a number of important industrial applications. It is particularly useful as an antirust coating on iron. In the USA upwards of 60 per cent of all cadmium is used for electroplating.[21] It gives better protection than does zinc and in the automobile and other industries cadmium plating has replaced galvanising of components. It was also used in quantity for plating food and beverage containers for some years, but this application was banned in the USA and Europe after the toxicity of cadmium was recognised. In the UK and Germany less cadmium is used for electroplating than in the USA and a higher proportion is used for pigments and as stabilisers in plastics.[22] Cadmium sulphide and cadmium sulphoselenide are very useful as pigments in paints and plastics, while cadmium stearate is used as a stabiliser in plastics.

Cadmium is found in a number of important alloys. It confers stiffness to copper and an increase in mechanical resistance at high temperatures. These copper–cadmium alloys are used in cooling devices such as car radiators. The metal is also used in some solders. A most important application is as an electrode component in alkaline accumulators.

CADMIUM IN FOOD AND BEVERAGES

Except where there has been pollution, cadmium is normally at quite a low concentration in foodstuffs. However, a wide range of concentrations has been reported by different workers. Some of these results possibly reflect defective analytical techniques rather than absolute values.

A 'market basket survey' carried out in Australia[23] in 1976 found a range of 0.095–0.987 mg/kg with a mean of 0.469 mg/kg of cadmium in food in daily family use. No samples had more than 1.0 mg/kg of the metal. Levels in individual foodstuffs were: bread, < 0.002–0.043; potatoes, < 0.002–0.051; cabbage, < 0.002–0.026; apples, < 0.002–0.019; poultry, < 0.002–0.069; minced beef, < 0.002–0.028; kidney (sheep), 0.013–2.000; prawns, 0.017–0.913 mg/kg.

The same pattern of low levels in vegetables and fruit, with higher amounts in meat and seafoods, has been reported for other countries also. Lisk[24] recorded a range of 0.05–3.66 mg/kg in seafoods and 0.19–3.49 mg/kg in meat. Lower levels have been reported from the UK, with < 0.01–0.09 mg/kg in meat and fish. The Australian survey of 628 samples of fish found a range of 0.01–0.2 mg/kg with a mean of 0.05 mg/kg.

The effects of industrial and other forms of contamination are seen in results for cadmium in rice from Japan, where grain from non-contaminated areas had 0.05–0.07 mg/kg, while from contaminated areas it contained approximately 1 mg/kg.

Cadmium levels in water, in the absence of contamination, is seldom above 1 μg/litre. Contamination may occur as a result of the use of galvanised pipes and cisterns. Cadmium-containing solders in water heaters and other fittings can be another cause. Pollution of water due to the use of cadmium-rich sewage sludge for agricultural purposes has been reported on a number of occasions.

Cadmium in cows' milk is generally less that 1 μg/litre, except in the case of animals fed on contaminated fodder. Evaporated milk contains about 0.04 mg/kg. Human milk contains 0.019 mg/litre.[25]

Other beverages may have high levels of cadmium, depending on conditions of storage and handling. Soft drinks from vending-machines which were implicated in cases of food poisoning due to the use of cadmium-plated components had upwards of 16 mg/litre of the metal. Levels of 18.1–37.6 mg/litre have been found in illicit spirits.[26] The use of equipment not designed for contact with foods and beverages in the construction of the stills was responsible for such levels of contamination.

Daily uptake of cadmium from food and beverages, in the absence of pollution, has been estimated in the UK to be between 10 and 30 μg from a 70 kg man.[27] In Sweden, levels of intake have been estimated to be between 10 and 20 μg/day.[28] Higher figures of 26–61 μg/day have been given for 15- to 20-year-old males in the USA, with an average daily intake of 39 μg.[29] Because of the known toxicity of cadmium and of the serious cases of contamination that occur from time to time, most countries monitor the levels of cadmium in the diet carefully. However, specific regulations restricting levels of the metal in foods are only enforced in a few countries. Reliance is placed on general regulations concerning heavy metals in food until such time as it is believed that more detailed controls are necessary. In the UK, for instance, the official view is that while there are no statutory limits for cadmium in food at present 'it is recommended that periodic analyses of certain foods should be carried out with the possibility of limits being imposed if necessary'.[30] The FAO/WHO provisional tolerable weekly intake of 0.3–0.4 mg is about 50 per cent higher than the estimated average intake for a UK adult.

There are a number of factors which may increase the intake of cadmium by the human body, and these must be taken into account when permitted daily levels of exposure are being considered. Cigarette smoking has been shown to result in a large increase in cadmium absorption by the body,[31] for example.

On the basis of a number of analytical studies, it has been estimated that cigarette smokers inhale 0.1–0.2 μg of cadmium with each cigarette smoked. It is believed that absorption of cadmium from the lungs is between 25 and 50 per cent of total intake, which means that for every 20 cigarettes smoked some 0.5–2 μg of cadmium are absorbed.

ABSORPTION AND METABOLISM OF CADMIUM BY THE HUMAN BODY

Only about 6 per cent of the cadmium in food is absorbed by the body.[32] However, there is evidence that many factors can affect the level of absorption. Simultaneous low levels of other minerals, such as calcium and iron, and of protein, in the diet may increase cadmium absorption.

Cadmium is transported in the blood mostly bound to a low molecular weight protein, probably metallothionein. In the liver and kidneys, the organs in which cadmium is mainly stored, it is also bound to this protein. Metallothionein has a molecular weight of about 6000. It is rich in sulphydryl groups and is capable of binding upwards of 11 per cent of cadmium and of zinc.[33] This dual affinity for the two metals probably accounts for the important relationship between zinc and cadmium, which will be discussed later.

Most of the cadmium absorbed by the body is retained. There is a very small excretion through the kidneys and in faeces. As a result of such efficient retention, the biological half-life of cadmium in the human body is very long, perhaps as much as forty years.[34] Even at low levels of exposure to environmental cadmium, the body accumulates metal all through life.

Newborn babies have very little cadmium in their tissues. By the time he has reached 50 years of age, an American, East German or Swedish male will have between 15 and 30 mg of cadmium in his body. The kidney cortex alone will have 25–50 mg/kg. A Japanese of the same age will probably have a higher cadmium body burden, with 100 mg/kg in kidney cortex. The levels will, in addition, be higher if the man is a smoker. It is of interest that the increase in cadmium in the renal cortex which normally occurs in humans with age is paralleled by an equimolar accumulation of zinc. This is related to the presence in the kidney of metallothionein. Attempts to assess levels of exposure to cadmium by measurement of the levels of the metal in body tissues have not as yet been very satisfactory. Blood levels may reflect more recent exposure, but probably do not reflect long-term kidney accumulation. Urinary concentrations have been found, in certain cases, to follow the accumulation in the kidneys.[35] It would seem that, in theory, cadmium in urine could be used as an index of the body burden but further work is needed to establish this relationship with certainty.[36]

EFFECTS OF CADMIUM INGESTION ON HEALTH

The ingestion of cadmium in food or in drink can cause symptoms of nausea, vomiting, abdominal cramp and headache within minutes of ingestion. In severe cases, diarrhoea and shock can also develop. About 15 mg/litre of cadmium in water and other beverages are sufficient to bring about such symptoms. Fruit juices, alcoholic beverages and other acidic fluids stored in plated containers, vending-machines and pottery vessels have often been the cause of such poisoning due to leaching of cadmium from the surfaces of the vessels. In fact, several cases of poisoning attributed to zinc leached from galvanised vessels were probably primarily cadmium poisonings.

Long-term ingestion of cadmium results in serious disease of the kidneys as well as of the bone. The most typical feature of chronic cadmium poisoning is kidney damage. The proximal tubules are damaged and renal reabsorption is affected. Low molecular weight proteins which are normally reabsorbed are excreted in increased amounts, resulting in tubular proteinuria. Other disturbances in reabsorption of amino acids and phosphorus may also folllow. Disturbances in renal handling of phosphorus and of calcium may cause resorption of minerals from bone, as will be mentioned in more detail later. Even after exposure to cadmium ceases, renal reabsorption disturbances persist and it would appear that the initial damage to the kidneys is irreversible. Other symptoms that have been observed following prolonged exposure to cadmium are anaemia and hypertension. Schroeder[37] has reported higher than average concentrations of cadmium in people dying of cardiovascular disease.

Apparently it is not simply higher levels of cadmium, but higher ratios of cadmium to zinc in the tissues, that are implicated.

Interaction between Cadmium and Zinc and Other Metals
There is evidence that cadmium toxicity may be lessened by the simultaneous ingestion of other metals. In animals, cobalt and selenium as well as zinc and zinc chelates have been shown to have this modifying effect. A

suggestion has been made that cadmium toxicity is due to the replacement of zinc in zinc-dependent enzymes and this may account for the special importance of the zinc—cadmium ratio in the diet. The protein metallo-thionein is involved in the transport of both cadmium and of zinc. There is evidence that the binding of cadmium to metallothionein is inversely related to the level of toxicity. Thus the low molecular weight protein acts both as a detoxification agent and as a store for cadmium.

The protective role of zinc against cadmium poisoning has been the subject of investigation. Schroeder[38] has commented on the problem of modern wheat milling technology which results in a decrease in the total zinc content and an increase in the relative proportion of cadmium to zinc in the flour.

The most dramatic and much publicised experience of the effects of cadmium poisoning are those relating to bone mineralisation. These are probably, in the main, secondary effects following renal damage brought about by cadmium. However, calcium metabolism may also be affected even before renal damage has occurred.

Where calcium deficiency already exists, defects in bone mineralisation can develop even at relatively low levels of dietary cadmium. This has been shown in the case of animals by a number of workers.[39]

Itai-itai disease is an osteomalacia resulting from a combination of dietary calcium deficiency and tubular dysfunction due to cadmium poisoning. The cadmium was ingested in rice which had been irrigated with cadmium-polluted water. Rice formed the bulk of the food of the affected persons, with very little meat or dairy products. The symptoms resulting from such a calcium-deficient, cadmium-enriched diet were initially pains in the back and legs. Pressure on bones, especially the long bones in the legs and the ribs, caused further pain. As the disease progressed, even slight bumps and knocks caused bone fractures. The name given to the disease, *itai-itai* (or ouch-ouch), expresses in a macabre manner the suffering of these unfortunate victims of industrial pollution. In addition to these symptoms, skeletal deformities also resulted, with a marked decrease in body height.

There is evidence that exposure to cadmium oxide vapour among industrial workers may result in a higher than normal level of incidence of cancer, especially of the prostate. This association has been recognised by the International Agency for Research on Cancer.[40] However, there is as yet no convincing evidence that ingestion of cadmium will have the same effect.

Both genetic and teratogenic effects have been attributed to dietary cadmium. However, the evidence is not clear. Chromosome aberrations have been shown in some *itai-itai* victims.[41] Congenital abnormalities were observed in mice litters after their dams had been exposed long-term to water containing 10 mg/litre.[42]

ANALYSIS OF FOODSTUFFS FOR CADMIUM

Pre-concentration is normally necessary for the determination of cadmium in foods, since the metal is present in most foods at very low levels of concentration. The Analytical Methods Subcommittee recommends[43] wet digestion using sulphuric acid and hydrogen peroxide. Dry ashing can result in low recoveries since cadmium is volatile at temperatures over 500°C. Iodocadmate ion is formed by extracting the digest with potassium iodide and an ion-exchange resin (Amberlite LA-2). Concentration of the cadmium may also be brought about by complexing with ammonium tetramethylene dithiocarbamate and extracting with 4-methylpentan-2-one (isobutylmethylketone).

A colorimetric method using dithiozone can be used for the determination of cadmium in food extracts. It is reported to be accurate and sensitive.[44]

The most commonly used method today is atomic absorption spectrophotometry. Using an air—acetylene flame accurate results are obtained, but flame conditions have to be carefully controlled. Flameless atomic absorption spectrophotometry allows estimation of as little as 5 µg/kg, but chemical interference, especially from sodium salts, may result in inaccuracies.

The use of anodic stripping voltametry for the determination of cadmium in food has been reported.[45] Recovery has been good and results are comparable to those obtained by atomic absorption spectrophotometry.[46] Neutron activation analysis is also employed with accuracy and dependability.[47] As a result of the development of new techniques and improved accuracy, it is becoming increasingly clear that a great deal of earlier work based on atomic absorption spectrophotometry and the less dependable flame emission photometry was unreliable due to inadequate methods. A full comparative study of methods of analysis of cadmium in food has recently been produced jointly by the WHO, the US Environmental Protection Agency and the Commission for European Affairs.[48]

7

The Toxic Metalloids: Arsenic, Antimony and Selenium

Arsenic has been traditionally associated with the poisoner and the forensic scientist. No one would take lightly the presence of this element in food or drink. Yet arsenic is almost universally distributed in plant and animal tissue, and daily intake, even in the absence of pollution, may be as much as 0.5 mg per person. Indeed, there is some evidence to indicate that arsenic may be essential for human life.

Selenium, a metalloid of the sulphur group, is similar in some ways to arsenic. It is highly toxic and its presence in food and drink has resulted in many incidents of poisoning. Yet it, too, is now believed to be essential for human life.

Antimony, unlike the other two metalloids, does not seem to be an essential element for humans, though it is very widely distributed in animal and plant tissues. It, too, is toxic and, as is the case with the other two metalloids, the level at which it may be present in foodstuffs sold to the public is controlled by food laws in many countries.

In the case of each of these three elements, the chemical form in which they occur governs the resulting level of toxicity.

ARSENIC

CHEMICAL AND PHYSICAL PROPERTIES

Arsenic has an atomic weight of 74.9 and is number 33 in the Periodic Table of the elements. Its density is 5.7. In its crystalline form it is grey in colour. It sublimes at its boiling-point of 613 °C and this property has been used for many years in Marsh's and other similar qualitative tests for the metal.

Oxidation states of arsenic are -3, 0 and 3.5. Its two most common inorganic compounds are the trioxide, As_2O_3 (white arsenic), and the pentoxide, As_2O_5 (arsenic oxide). Other important compounds are arsenic trichloride and the various arsenates such as lead arsenate, copper aceto-arsenate and the gaseous hydride, arsine (AsH_3). Organic compounds of interest are arsanilic acid, dimethylarsanilic acid and cacodylic acid.

The toxicities of compounds of arsenic have been classified by Vallee[1] in the following order:

1. Inorganic arsenicals (trivalent arsenicals, arsenic trioxide, arsenite salts, pentavalent arsenate salts).
2. Organic arsenicals.
3. Arsine.

PRODUCTION

Arsenic is widely distributed in the environment. It occurs in almost all soils. However, no major ore of arsenic is known and the compound is obtained as a by-product of the smelting of copper, lead and some other metals. The arsenic, which usually is in the form of the trioxide in the natural state, is vaporised during smelting. The vapour travels in the flue gases to electrostatic precipators, where it is collected. The mixed dusts are removed and roasted to vaporise the arsenic once more. This is condensed in almost pure form in a cooling chamber or 'arsenic kitchen'. The product consists of about 97 per cent arsenic trioxide and 3 per cent of oxides of other elements. The most important of these is antimony trioxide.[2]

World production is about 50 000 tonnes each year. The rate of increase in use over recent years has been about 25 per cent per decade.

USES

Arsenic has only limited use in the metallurgical industry, though in its elemental form it is used in small amounts in the production of some alloys. It has the property of increasing hardness and heat resistance in steel.

The principal use of arsenic today is in agriculture and related areas. Herbicides, fungicides, wood preservatives of various kinds, insecticides, rodenticides and sheep-dips may all contain arsenic. The most important agricultural chemicals of this type include lead arsenate, copper arsenate, copper aceto-arsenite ('Paris green'), sodium arsenate ('Wolman salts') and cacodylic acid. It is the use of such chemicals in agriculture that has given rise to considerable concern about their forward transmission to the human consumer. This concern has resulted in a reduction in use or total banning of some of them. Lead arsenate, which is used as an insecticide on tobacco, can contribute to higher than average arsenic intake by smokers. Similarly the application of arsenical insecticides to apple trees can result in contaminated fruit and cider. Several cases of stock and even human poisoning have resulted from the use of arsenical sheep-dip. Other agricultural uses of arsenic compounds include the incorporation of arsanilic acid into pig and poultry feed as a growth-promoting agent.

The chemical industry uses arsenic in the manufacture of dyestuffs as well as of glass and enamels. In former times arsenic was widely used medicinally. Until relatively recently a preparation of the inorganic element known as Fowler's solution was used for the treatment of dermatoses and even as a tonic for the treatment of anaemia. A report [3] published in 1965 on several hundred patients who had been treated for extended periods, sometimes up to several years, with Fowler's solution showed that a high proportion had developed skin cancer. Organic arsenical compounds such as Salvarsan have been extensively used for the treatment of syphilis. It was possibly the availability of such pharmaceutical compounds that made arsenic, especially in the form of the trioxide, one of the most common homicidal poisons for many centuries.

ARSENIC IN FOOD AND BEVERAGES

Because of its wide distribution in the environment and, to some extent, due to its use in agriculture, arsenic is present in most human foods. However, the amount present is usually quite small—less than 0.5 mg/kg—and rarely exceeds 1 mg/kg, except in the case of certain marine organisms which can concentrate the element. Soils normally contain between 1 and 40 mg/kg in the absence of industrial or agricultural contamination. Uptake by plants has been found to depend not only on soil concentration of the element but also on the species of plant as well as on the nature of the soil.

The use of phosphate fertilisers may contribute to a higher than normal soil content of arsenic. The phosphate rock used in the manufacture of such fertilisers, as well as of some detergents, may contain not inconsiderable amounts of arsenic.[1,4]

The range of arsenic levels in foodstuffs in the absence of serious pollution is: cereals, 0–2.4 mg/kg; fruit, 0–0.17 mg/kg; vegetables, 0–1.3 mg/kg; meats, 0–1.4 mg/kg; dairy products, 0–0.23 mg/kg. Seafoods generally contain more arsenic, with levels of 1.5–15.3 mg/kg[5].

Arsenic is detectable in almost all potable waters. However, there can be considerable differences in levels in drinking-water from different areas, depending largely on the nature of the underlying rock. In most parts of the world the average content of river water is approximately 0.5 μg/litre with a range of zero to approximately 0.2 mg/litre.[6] US federal regulations for drinking-water set a maximum of 0.01 mg/litre of arsenic.[7] Particular water sources, such as those from hot springs and spa waters, very often exceed such standards, with levels from about 0.5 to 1.3 mg/litre. Regular use of such waters for domestic purposes can result in an excessive intake and produce symptoms of chronic arsenic toxicity. There are several other natural and man-made causes of increased arsenic intake in the diet.

Fish, especially crustaceans, are, as has been noted, accumulators of many potentially toxic elements. Prawns from UK coastal waters have been found to contain as much as 170 mg/kg of arsenic. Shrimps from southern US coastal waters had 40 mg/kg. Free-swimming fish do not normally contain so much, though people consuming large amounts of fish as a normal part of the diet accumulate more arsenic than those in non-fish-eating areas.

Excessive intake of arsenic due to naturally occurring high levels in drinking-water has been observed in several parts of the world. 'Regional endemic chronic arsenicism' caused by drinking water containing between approximately 1 and 4 mg of As_2O_3 per litre has been reported in the Argentine.[8] A similar situation has been observed in Chile.[9] The ingestion of well-water containing 0.6 mg/litre of arsenic has led to localised chronic arsenic poisoning in Taiwan.[10]

Industrial pollution and accidental contamination can also result in higher than normal levels of arsenic in foodstuffs and beverages. Studies conducted in the area surrounding a coal-burning power-station in Czechoslovakia showed considerable pollution of the environment with arsenic. Coal used in the USA contains an average of 5 g of arsenic per tonne. However, that used at the Novaky power-station in Czechoslovakia contained 200 times as much. The plant emitted up to one tonne of arsenic from its flues each day. Drinking-water in the surrounding area was found to have 0.07 mg/litre of arsenic while in surface water it reached a level of 0.21 mg/litre.[8] Chronic arsenic poisoning has been alleged to have occurred in cattle near coal-fired brick kilns in the UK.[11]

More than 10 tons of arsenic a day were emitted by a Swedish smelter in the 1940s and reports of similar local contamination have appeared in the USA.[12]

Accidental contamination of foodstuffs with arsenic or arsenic-containing additives has resulted in a number of serious cases of poisoning. Starch which had been hydrolysed with arsenic-contaminated acid and which was used in beer fermentation caused arsenic poisoning of several thousands of people in the UK in a notorious incident in the late nineteenth century.[13] An even more serious case occurred in Japan in 1955 when more than 12 000 infants who were bottle-fed on a formula containing dried milk which had been contaminated with arsenic trioxide were poisoned. The arsenic was introduced accidentally in sodium phosphate which was used to stabilise the milk powder. The sodium phosphate was a waste product generated during the refining of aluminium from bauxite and it unsuspectedly contained a substantial amount of arsenic. More than 120 of the infants died after ingesting the poisonous mixture over 33 days at a rate of about 3.5 mg of the trioxide each day.[14]

The use of arsenicals as pesticides in vineyards has resulted in some cases of poisoning. Cirrhosis of the liver has been traced to the consumption of arsenic-contaminated wine.[15] Because of such dangers resulting from the horticultural use of arsenic and its compounds, some countries have set tolerance limits to such residues in foodstuffs. In the USA, for instance, this is 3.5 mg/kg (as As_2O_3) in vegetables.[16]

The average daily intake of arsenic in the diet will obviously depend on the type of food eaten. However, in the absence of severe environmental pollution and where seafoods do not make up a significant part of the diet, daily intake has been calculated by Schroeder and Balassa to be in the region of 0.2 mg.[4] Similar levels of intake have been reported for other technologically developed countries.

Both trivalent and pentavalent arsenic are easily absorbed from the gastro-intestinal tract. Actual amounts absorbed seem to depend on the chemical form of the compound involved and the composition of the whole diet. Information on the actual fraction of ingested arsenic which is taken into the bloodstream is lacking. Once absorbed, arsenic is quickly distributed to all organs and tissues as a protein complex, probably with a-globulins. After about 24 hours, concentrations in most organs appear to decrease, but accumulation in skin may increase for several days after ingestion.[17] Preferential accumulation in skin, nails and hair and similar tissue has been noted.[18] There is some accumulation in bone and muscle also, and even though concentrations in these tissues may be low, overall amounts may be high because of the mass of the tissues. Along with skin, bone and muscle represent the major arsenic depots of the body.

Total adult body content has been reported from the USA to be between 14 and 20 mg.[4] In a series of analyses of tissues and organs from healthy persons who died of accidents in Scotland, levels of arsenic in different parts of the body were found to be as follows:[19] brain, 0.012 mg/kg; whole blood, 0.036 mg/kg; heart, 0.021 mg/kg; bone, 0.053 mg/kg; muscle, 0.062 mg/kg; skin, 0.080 mg/kg; teeth, 0.049 mg/kg; nails, 0.283 mg/kg; hair, 0.460 mg/kg.

In general, biological half-time is short. Animal experiments give values of 36–60 hours. In the case of a person who drank contaminated wine, the half-time for trivalent arsenic was 10 hours and for methylated forms 30 hours.[20]

Arsenic excretion takes place mainly in urine, with a little in faeces. Concentrations in urine have been used as an index of exposure, but the relationship is not straightforward. The total dietary pattern as well as the chemical form of arsenic should be considered when levels in urine are being used as a measure of intake.

Arsenic in blood is difficult to measure accurately and wide variations in blood concentrations and range have been reported. Levels of arsenic in hair have been reported to reflect atmospheric pollution. Such measurements have also been used to study ingestion of the element. Normal hair concentrations, in the absence of excessive consumption of arsenic, have been variously given as less than 1 mg/kg,[21] 0.02–8.17 mg/kg[22] and below 4 mg/kg.[23] Levels of as high as 85 mg/kg in hair of persons chronically poisoned by drinking arsenic-contaminated well-water over a prolonged period have been reported. In cases of acute arsenic poisoning, levels ranging from 5 to 700 mg/kg of hair were found.[23]

METABOLISM AND TOXIC EFFECTS OF ARSENIC

Arsenic can cause both acute and chronic poisoning. The former is well known to the forensic scientist and though, apparently, no longer commonly used in suicide or homicide, poisoning does occur fairly frequently as a result of accidental ingestion of arsenic-contaminated food. Arsenic trioxide is the most common chemical form met with in such accidents. A fatal dose is of the order of 70–180 mg of the trioxide.

The pentavalent compound is less toxic than the trivalent form. Arsenic is a general protoplasmic poison. It binds organic sulphydryl groups and thus inhibits the action of enzymes, especially those concerned with cellular metabolism and respiration. The principal pharmacological effect is dilation and increased permeability of the capillaries, especially in the intestines.

Chronic poisoning by arsenic results in loss of appetite and subsequently of weight, gastro-intestinal disturbances, peripheral neuritis, conjunctivitis, hyperkeratosis and melanosis of the skin. This latter effect on skin pigmentation is characteristic of prolonged exposure to the element and may be related to the development of skin cancer. The connection between exposure to arsenic and the development of cancer, especially of the skin, has been postulated by a number of workers. Neubauer[24] has recorded the finding of unusually high levels of incidence of skin cancer in areas of the UK where arsenic was present in drinking-water at 12 mg/litre. Lung cancer, at least among workers exposed to arsenic trioxide, has also been found. Claims have also been made that high levels of arsenic in the diet can lead to cancer of organs other than lungs and skin, but this has not been confirmed.[25] Teratogenic effects in animals after administration of arsenicals have been reported[26] but evidence of similar effects in humans is lacking.

Interaction between Arsenic and Other Metals
Arsenic has been found to counteract the toxic effects of selenium.[27] It has been added to poultry and cattle feeds to suppress toxicity in areas of high natural selenium content. Similarly, sodium selenite injected simultaneously with sodium arsenate, which is teratogenic in animals, prevented development of malformation.[28] Interactions betwen arsenic and cadmium have also been shown. No doubt, common modes of action of these metals on biological tissue, especially through interaction with protein sulphydryl groups, account for some of the observed results.

Arsenic an Essential Human Nutrient?
Though arsenic is certainly toxic, it is present in almost all foods and in the human body, at least at a low level of concentration. It has been shown to have beneficial effects on animals, apart from its ability to counteract selenium toxicity. The use of organic arsenicals to improve the growth, health and feed efficiency of poultry and pigs is now thoroughly established.[29] Reports have been published of the effects of arsenic in improving the appearance of the skin and hair of various animals.[4] The level of intake at which such improvement was observed was minute—as little as 1 μg/day in the case of rats. There is as yet no convincing evidence that arsenic is also beneficial to humans, but the possibility cannot be ruled out.

ANALYTICAL METHODS FOR ARSENIC

The traditional tests for arsenic, favourites of the detectives of fiction as well as of forensic scientists of more recent years, were qualitative rather than quantitative. The methods of Marsh, Gutzeit and Reinsch all gave visual evidence, and conclusions as to quantity had to be very cautiously made.

Colorimetric methods of analysis which allow quantitative determinations in the range of 1–50 mg/litre can be carried out using silver diethyldithiocarbamate.[30]

A widely used and convenient method of analysis for arsenic is atomic absorption spectrophotometry applied to arsine gas generated from the arsenic.[31] Commercially produced arsine-generating kits are available and are used in conjunction with standard equipment. An acetylene–nitrous oxide flame is recommended. Interference occurs due to molecular absorption of flame gases and solution species at the extreme ultraviolet region of the spectrum, where the most sensitive lines for arsenic are found. This non-atomic absorption can be estimated by means of a continuum light source such as a hydrogen lamp, and background corrections made.

Neutron activation analysis has been used with success for the determination of arsenic at levels down to the microgram range. The method has been applied to the accurate determination of arsenic in samples as small as a single strand of hair.[32]

It is often of interest to determine the exact chemical form in which arsenic is present. Pulse polarography has been used to distinguish trivalent from pentavalent arsenic[33] in water. Gas chromatography has been used to distinguish various organic compounds of arsenic from inorganic forms.[34]

ANTIMONY

CHEMICAL AND PHYSICAL PROPERTIES

Antimony, which has been given the chemical symbol Sb after its classical name of *stibium*, has an atomic weight of 121.8 and is number 51 in the Periodic Table of the elements. It is a fairly heavy element with a density of 6.7. Though related closely chemically to arsenic, its physical properties are more metallic than non-metallic. Its melting-point is 631°C and the liquid vaporises at 1750°C. Its crystalline form is a silver-white metal, hexagonal in shape. Its oxidation states are 3 and 5. Among the most important of its compounds are the trioxide and pentoxide, trichloride, trisulphide and pentasulphide, antimony potassium tartrate and the gas stibine (the trihydride, SbH_3). Several organic compounds are also known.

PRODUCTION

An important ore of antimony is the sulphide stibnite. Several other ores also occur. About 70 000 tonnes are produced each year, principally in South Africa, Bolivia and China.

USES

Antimony is used in the manufacture of many alloys, especially with lead and copper. These are employed for making bearings, battery parts, printers' type, solder, ammunition, castings, cable coverings and several other products. It is also used extensively in the chemical industry to produce fireproofing chemicals, as well as in paints and lacquers, rubber, glazes and pigments for ceramic and glass manufacture. A certain amount is employed in the pharmaceutical industry, particularly in preparations for the treatment of parasitic infestations. One of these, widely used in tropical regions, is tartar emetic.

ANTIMONY IN FOOD AND BEVERAGES

Little is known about daily intake of the element or even about normal levels in food and drink. Figures from the USA indicate an intake of 0.25–1.25 mg day for children.[35] Much lower levels of intake have been reported from Sweden.[36] Figures for individual foodstuffs are lacking. Concentrations in water have been given as approximately 0.1–02 μg litre for both river and marine samples.[37, 38] No official limit has been set for the presence of antimony in drinking-water. The US Environmental Protection Agency has recommended a limit normally not exceeding 0.1 mg litre, with the further proviso that over long periods levels should not exceed 0.01 mg/litre.[39]

Reports have been published[40] of high levels of antimony in certain foodstuffs due to contamination. This has frequently been the result of the preparation or storage of food in containers glazed with an antimony-containing enamel. Monier-Williams[41] found that as much as 100 mg/litre of antimony could be dissolved by a 1 per cent citric acid solution from enamelware. A case of poisoning involving 56 persons who had drunk lemonade made from crystals and prepared in a white enamel bucket has occurred.

Several countries have established maximum permissible levels of antimony in beverages and other foodstuffs. Both Australia and New Zealand set the limit at 0.15 mg/litre for beverages, but while the former country permits 1.5 mg/kg for other foodstuffs, New Zealand has a lower limit at 1.0 mg/kg. The UK has no general specified or recommended limits except in the case of food colourings, where a maximum of 100 mg/kg is laid down.

METABOLISM AND BIOLOGICAL EFFECTS OF ANTIMONY

Not very much is known about the uptake and behaviour of antimony in the human body. From the results of animal experiments it would appear that about 15 per cent of ingested antimony is absorbed in the gut. The element is concentrated in organs such as liver and kidneys as well as skin and the adrenals. Total body burden in the normal human being is about 1 mg.[42] Excretion is reported to be rapid initially but there may also be a long-term component.[43]

There is no evidence that antimony is an essential trace element for humans or other animals. Most of its effects on humans are known from industrial exposure involving inhalation of dust and fumes. The few cases of poisoning by ingestion which have been reported resulted mainly from drinking antimony-containing soft drinks which had been made or stored in enamelled containers. The symptoms of poisoning are colic, violent nausea, weakness and collapse with slow or irregular respiration and a lowered body temperature. Poisoning has also resulted from the use of antimony compounds in the treatment of parasitic infestations. Nausea and vomiting have been commonly reported during such treatment, and a few cases of sudden death.[44]

ANALYSIS OF FOODSTUFFS FOR ANTIMONY

The method most frequently used today is atomic absorption spectrophotometry. Detection limits are in the region of 0.5 mg/litre in aqueous solution.[45]

Hydride generation with flame atomic fluorescence spectrophotometry has recently been developed with an improved detection limit.[46] Neutron activation is also used where facilities are available. Colorimetric methods such as that using rhodamine B are still frequently mentioned in the literature.[47]

SELENIUM

CHEMICAL AND PHYSICAL PROPERTIES

Selenium occurs in Group VI of the Periodic Table, along with sulphur, tellurium and polonium. These elements make up the sulphur family. It has an atomic weight of 78.96 and is element number 34 in the Table. Its density is 4.79. Selenium occurs in a number of allotropic forms, one of which is metallic or grey selenium. This crystalline form contains parallel 'zigzag' chains of atoms and is the stable form at room temperature. Its electrical conductivity is increased on exposure to light, a property of considerable importance in photoelectric cells. Another crystalline form is the sulphur-like monoclinic or red selenium. Selenium also exists in red and black amorphous forms.

Chemically, selenium has many properties similar to those of sulphur, though it is a weaker oxidising and reducing agent than the more electronegative sulphur. It burns in oxygen to form the solid dioxide, SeO_2. It also combines directly with the halogens and with many metals and non-metals. Selenides of several metals can be prepared by direct reaction of the elements. Hydrogen selenide, H_2Se, is formed when dilute acids react with some metallic selenides. It is a colourless gas with an offensive odour and is highly toxic.

Selenous acid, H_2SeO_3, is produced by the reaction of the dioxide with water. A series of selenites and hydrogen selenites, corresponding to the sulphites and hydrogen sulphites, occurs. Corresponding to the reactions of sulphur trioxide, selenium trioxide, SeO_3, a white hygroscopic solid, forms selenic acid, H_2SeO_4, which is very similar in properties to sulphuric acid. A series of selenates also exists.

PRODUCTION AND USES

Selenium is not abundant, though it is a widely distributed element in the earth's crust. On average, soil contains about 0.2 mg/kg,[48] though pastures which are much poorer, as well as others which contain much more of the element, are not uncommon and cause problems in agriculture. Selenium rarely occurs in native form, but usually as selenides. These are often associated with the far more abundant sulphur. Selenium can be extracted from flue dusts produced during the combustion of metallic sulphide ores such as pyrites, FeS_2. It is also obtained from residues in lead chambers used in the production of sulphuric acid. Most of the world's supply comes from Canada, the USA and Zambia, with the latter producing only a very small quantity. The element has a number of important specialist uses, one of which has been mentioned already. This is based on the element's ability to conduct electricity more easily when exposed to light. A small electric current is also generated in the selenium by light. These properties have been used in the construction of photoelectric cells for light meters, automatically operating switches and locks, solar batteries, Xerox and similar copying machines and other electronic devices. Selenium of high purity is also used as an efficient rectifier for alternating currents.

Apart from its use in the electronic industry, selenium is employed in the production of certain stainless steels and special copper alloys in order to improve their machinability. It is used in glass manufacture to reduce the green colour which results from the presence of iron silicates. On its own in larger amounts selenium imparts a red colour to glass and ceramics. Along with sulphur, selenium is used to vulcanise rubber. The element also, to a small extent, has an agricultural use as a feed supplement and as a top-dressing for pastures when levels are naturally low in the environment.

SELENIUM IN FOOD AND BEVERAGES

Until recently, information on the levels of selenium in foods and the daily dietary intake was not readily available. However, since it has been recognised that the element, which had long been known as a toxic contaminant of animal and human food, was also an essential nutrient for man, several surveys of levels in foodstuffs have been carried out. A recent study in the UK[49] found the following range for different groups of commodities: cereals, 0.03–0.23 mg/kg; meat and fish, 0.09–0.16 mg/kg; fish alone, 0.28–0.38 mg/kg; milk, <0.01–0.01 mg/kg; vegetables, <0.01–0.02 mg/kg; fruit, <0.01 mg/kg of fresh food. The average intake per person (adult) per day was calculated to be 60 μg. The UK figures are low compared to those reported from some other countries. This has been attributed to the generally low environmental level of selenium in Britain.

New Zealand, where much of the farmed soil is also poor in selenium, has an even lower daily intake per head of 25 μg.[50] Average adult daily intake for the USA is 60–150 μg,[51] Canada 110–220 μg,[52] Netherlands 110 μg, France 166 μg and Italy 141 μg.[53]

It is of considerable interest and significance that while several plant accumulators of selenium are known, no vegetable used for human consumption seems, fortunately, to have this ability. The pasture plant, *Astragalus racemosus*, has been found to accumulate 15 g/kg on selenium-rich soil in the US mid-West. An annual legume, *Neptunia amplexicaulis*, growing on seleniferous soil in Queensland, Australia, had over 4 g/kg (dry weight). Both plants can contain more than enough selenium to poison grazing animals.[54]

Among all the foods investigated in different countries, fish, some meat and nuts appear to be the richest sources of the element. Thus the dietary pattern can have considerable effects on the level of daily intake of the element. Canadian sea fish, for instance, contain 0.9 mg/kg,[55] substantially higher than the levels reported in the UK.

Levels found in various fish products in Norway[56] are also high, ranging from 0.19 mg/kg for fish balls to 4.43 mg/kg for lobster roe.

Nuts are another rich source of selenium, a fact which may be of significance for those committed to a vegetarian diet. The UK study[49] found individual samples of Brazil nuts which contained as much as 53 mg/kg of the element. Cashew nuts, walnuts and peanuts were also relatively rich in selenium, though not at the levels found in Brazil nuts.

Water supplies do not constitute a significant source of selenium for man, even in areas where environmental levels of the element are high.

A recommended daily intake of 60–120 μg of selenium has been suggested by the US National Research Council. The Council also indicated that toxicity will occur after prolonged ingestion of upwards of 3000 μg/day.[57] No recommendation has yet been made regarding requirements or tolerable levels by the WHO.[58]

METABOLISM AND BIOLOGICAL EFFECTS OF SELENIUM

Gastro-intestinal absorption of selenium is related both to the chemical form of the element occurring in food and the amount ingested. At high levels, sufficient to be toxic to animals, selenium is rapidly absorbed from selenium-rich plants and also from food to which soluble salts have been added. Elemental selenium itself is poorly absorbed. Absorption of selenium in animals appears to be mainly from the duodenum. The element is carried in the plasma, probably bound to proteins, to all tissues. It appears that selenium may replace sulphur in proteins and other cellular components and form, for example, selenocystine and selenomethionine. Absorbed selenium is rapidly eliminated once dietary intake is reduced. Excretion is by faeces, urine and expired air.

Selenium metabolism can be affected by the presence of sulphate and other substances. In animals both an ameliorating effect of sulphate on selenium toxicity and an exacerbating effect on selenium deficiency have been shown.[59] The most effective dietary factor in reducing the toxicity of selenium is arsenic.[60]

The discovery in the late 1950s that selenium was an essential nutrient for animals was a considerable surprise.[61] It had long been recognised as a highly toxic element by agricultural scientists. Two diseases of stock known as 'alkali disease' and 'blind staggers' were known to occur in the Great Plains of North America and in other parts of the world. These diseases were identified as symptoms of acute and chronic selenium poisoning. Stock which fed on pasture growing on selenium-rich soils were found to accumulate toxic levels of the element in tissues. Some symptoms of selenium toxicity have also been noted in humans living in such seleniferous areas.[49] In some countries maximum permitted levels of selenium in food are being established. For example, it is proposed that permitted levels in Australia will be as follows: beverages and liquid foods, 0.2 mg/kg; edible offal, 2.0 mg/kg; all other foods, 1.0 mg/kg.[62]

The mechanism of selenium toxicity in animals is not clear. There is some evidence that the element interferes with processes of cellular oxidation by replacing sulphydryl groups in dehydrogenating enzymes.

That selenium was not only a toxic element but also a trace nutrient was shown by the work of Schwartz and others.[54] As little as 0.1 mg/kg of selenium in feed was found to prevent muscular dystrophy ('white muscle disease') in cattle and sheep, exudative diathesis in poultry and hepatosis dietetica in pigs. Selenium appears to be required for a number of important enzymes in animals as well as man. These enzymes are selenoproteins and include glutathione peroxidase in human cells. Glutathione peroxidase is a principal protective agent against accumulation of hydrogen peroxide and organic peroxides in cells and tissues. A number of other enzymes have also been shown to be selenoproteins. Other biochemical roles have been proposed for selenium. It may help to protect animals against mercury toxicity. There is evidence for a close relation between selenium and vitamin E. Lack of either causes muscular dystrophy in many animals. It has been suggested that vitamin E protects reduced selenium from oxidation from the selenide, Se^{2-}, which is its effective state in cells. There is some epidemiological evidence that a high intake of selenium during childhood will result in an increased incidence of dental caries.[63] However, results of other studies do not entirely support this view. Selenium has also been implicated both as a carcinogen in rats and as a protective agent against cancer in man. In niether case is the evidence entirely convincing.[54]

ANALYTICAL METHODS FOR SELENIUM

Several methods can be used for the estimation of selenium in foods. Atomic absorption spectrophotometry has been employed effectively by a number of workers, for example for analysis of fish and fish products.[56] However, strong absorption by non-atomic species in the flame occurs and background correction using a hydrogen continuum lamp is necessary. Chemical interference also occurs from some elements, especially barium, lead, lithium, sodium and strontium. Use of a nitrous oxide–acetylene flame will reduce this interference but with loss of sensitivity.

Greater sensitivity is obtained by using the vapour generation method. In this, selenium compounds in digested food samples are reduced to the volatile hydride and passed into a special quartz cell where absorption in a flame is measured.

A fluorimetric method for selenium in a variety of different foodstuffs was used in the UK survey mentioned above.[49] Food samples were digested with nitric and perchloric acids. Hydrogen peroxide was added to the digest to ensure the reduction of selenium to the tetravalent state. The selenium was then complexed with 2,3-diaminonaphthalene, extracted into cyclohexane, and estimated fluorimetrically at 518 nm. The procedure permitted recoveries of some 95 per cent, with good reproducibility and accuracy of results.

8

The Packaging Metals: Aluminium and Tin

ALUMINIUM

Aluminium is the most common metal in the lithosphere, making up some 8 per cent of the total crust of the earth. It occurs usually in the form of silicates. In spite of its abundance, it was not until man developed high technological capability and had an abundance of electrical energy that he was able to extract the metal from these readily available but intractable sources. The metal was originally isolated in 1825. It had its first major public appearance at the Paris Exhibition in 1855. Even after the introduction of the electrolytic method of extraction, it was not at first considered to be a very useful addition to industry. However, once further progress in metallurgical techniques allowed casting, rolling and joining of aluminium, its special properties recommended it for a wide variety of applications in various branches of industry. Its resistance to corrosion and its lack of toxicity were particularly appreciated, and aluminium cooking utensils and, later, cans and other food containers quickly came into use. Of only minor significance in the 1930s, when less than 10 000 tonnes were produced, it is today of major importance, used in a wide variety of important applications.[1] The growth rate of aluminium production in recent years has been three times the combined growth rate of all other metals. It is predicted that total use in the USA will be between 20 and 40 million tonnes by the year 2000.[2]

CHEMICAL AND PHYSICAL PROPERTIES

Aluminium has an atomic weight of 27 and is number 13 in the Periodic Table of the elements. It is a light metal, with a density of 2.7. Its melting-point is 660.4 °C. It is soft and ductile, silver-white in colour, with good electrical and heat conductivity. It has an oxidation state of 3. It is extremely resistant to corrosion, though its alloys are less so. Its principal inorganic compounds are the oxide, hydroxide, sulphate, fluoride and chloride. It also forms organic compounds, some of which are highly reactive and become hot and fume on exposure to air.

PRODUCTION

Aluminium is present in clays and rocks as alumino-silicates and alumino-hydroxides and is universally distributed in soils. However, the chief commercial sources are the ores bauxite (the hydrated aluminium oxide) and cryolite (sodium aluminium fluoride). Major deposits are found in the Caribbean region, Brazil and South Africa. To extract the metal from bauxite, the ore is converted to aluminium hydroxide by treatment with caustic soda. This is then converted to the oxide by calcination, and aluminium is produced from this by electrolysis in smelted cryolite at about 970 °C.

A source of aluminium of particular interest to the food and beverage industries is bentonite, a naturally occurring hydrated colloidal alumino-silicate. This is used, for example, in beer-making as a clarifying agent.

USES

Because of its many valuable properties, such as lightness, electrical conductivity and corrosion resistance, uses of aluminium exceed those of any other primary metal. About half of the world's production is used in the electrical industry, where copper is being replaced in many applications by aluminium. A considerable amount is also used in motor vehicle engineering, aircraft and building construction. Household appliances and utensils are now frequently made of the metal, as are food-processing vessels and large catering equipment. One of the major uses, and one which is rapidly expanding, is in the packaging industry in foil, cans and other food and beverage containers.

In compound form aluminium is used in the food industry, as a food additive and in other ways. Particle sedimentation in water treatment uses aluminium sulphate. Aluminium hydroxide, or alumina, finds wide application ranging from high temperature-resistant refractories, ceramics, whiteware, toothpaste, food additives such as baking powder and many others. Other aluminium compounds are also important as pharmaceutical and therapeutic agents. They are used, for example, to prevent hyperphosphataemia in renal disease, as an antacid, as an adjuvant for vaccines, as an antiperspirant and in several other ways.

ALUMINIUM IN FOOD AND BEVERAGES

In spite of considerable interest ever since aluminium first came into general use for making household and food-processing equipment, only a limited number of reports on the levels of the metal found in food or consumed in the daily diet have been published. Total daily intake is estimated to be about 80 mg.[2] A total diet study for 16- to 19-year-old boys in the USA showed a daily intake of 8.8–51.6 mg of aluminium.[3] Unprocessed foods contain about 10 mg/kg of the metal, though certain vegetables may accumulate levels of up to 150 mg/kg. Spices and root crops, probably due to soil contamination, are particularly high in aluminium. The use of aluminium compounds in processing can increase levels of aluminium in food considerably. For example, processed cheese was found to contain 695 mg/kg. Baking powder may contain over 20 000 mg/kg. Some candies and chewing-gum contain up to 100 mg/kg of aluminium.

There is little evidence that cooking food in aluminium vessels leads to an increase in intake of the metal. Under acid or strongly alkaline conditions uptake by food is only slightly higher than when conditions are neutral.[4] In the absence of pollution, fresh water may contain about 10 mg/litre, though domestic supplies where aluminium sulphate is used at the treatment plant may contain higher levels. Levels in cows' (as well as in human) milk have been reported to be in the range of 1–2 mg/litre.[5]

ABSORPTION AND METABOLISM OF ALUMINIUM
BY THE HUMAN BODY

Most of our information on aluminium absorption and metabolism comes from studies on persons receiving aluminium hydroxide treatment for hyperphosphataemia in renal failure. Aluminium forms an insoluble compound with phosphorus in the gastro-intestinal tract and this prevents excess phosphate absorption. Results show that in these cases about 15 per cent absorption of aluminium takes place.[6]

Normal tissue and all organs of the human body that have been analysed have been found to contain aluminium, indicating that absorption and distribution within the body normally takes place. Rats fed on a diet containing high levels of aluminium showed increased levels in liver, brain, testes and blood.[7]

The bulk of aluminium excreted from the body appears in the faeces. Urinary excretion is very limited.

EFFECTS OF ALUMINIUM INGESTION ON HEALTH

As long ago as 1886, when it had begun to be used in the construction of cooking utensils, claims were made that ingestion of aluminium produced toxic effects. Considerable controversy resulted and continued for many years.[8] Forty years later Spira resurrected the matter and presented further evidence which he said substantiated the earlier claims.[9] Other reports and studies have since been published both in support of and in contradiction to the view that the use of aluminium cooking utensils was hazardous and could result in food poisoning. In view of the uncertainty that existed and to reassure the public, the American Medical Association Council on Food and Nutrition felt that it should make a definite statement on the matter.

The Council considered that evidence was lacking to support the view that the small amounts of aluminium ingested in the diet, even when food was cooked in aluminium utensils, had any ill effect on health.

Though the controversy is occasionally resurrected today, no convincing evidence has emerged in the intervening years since to disprove the American Medical Association's declaration.[10] Nevertheless, the fact that the use of utensils made of this metal can result in discoloration of foodstuffs containing certain natural pigments still causes concern to some. Moreover, evidence that in some cases of pharmaceutical use toxic side-effects have arisen has raised doubts in the minds of some about its safety. However, though toxicologists keep the matter under review, it seems highly unlikely that, under normal conditions of life, aluminium, when used in the manufacture of containers for food or drink or catering and food-processing equipment, presents a hazard to human health.

ANALYSIS OF FOODSTUFFS FOR ALUMINIUM

Problems due to interference by other substances, and also by contamination resulting from the widespread distribution of aluminium in the environment, make analysis of foodstuffs for aluminium by colorimetric and some other methods difficult. No specific reagent for aluminium exists and separation and concentration of samples present particular difficulties. Thus procedures in which pre-treatment of samples can be kept to a minimum are highly desirable. Emission spectroscopy has been shown to be especially suitable from this point of view and is a specific and sensitive analytical technique.[11]

Atomic absorption spectrophotometry is a widely used method for analysis of aluminium in food and other biological materials. However, the method requires care to overcome interference from other elements, in particular the alkali metals as well as nickel, titanium, iron, chromium and manganese. Addition of a readily ionisable element such as potassium helps to overcome enhancement interference from these elements. A fuel-rich flame as well as pre-treatment of samples to remove silicon,[12] which also causes problems by depressing absorption, improve the method. A method for analysis of aluminium in whole blood using atomic absorption, which is specific and sensitive, has been reported by Langmyhr and Tsalev.[13]

Neutron activation analysis is the most sensitive analytical technique available for aluminium. In some cases sample pre-preparation is not required, which is a considerable advantage. However, in food samples with a high phosphorus content a preliminary separation step is necessary.[14]

TIN

Tin is one of the ancient metals. Long before the Romans crossed from Gaul to Britain, Phoenician merchant ships had been calling at Cornish ports to load the rare and valuable metal that the Celtic miners had won from their underground galleries and pits. The product of those and other ancient mines was used mainly as an alloy in combination with copper to produce bronze. It was also alloyed with lead to give pewter. Only in recent times has tin been used in its own right on a large scale and then only in the thinnest possible layer, as tin plate on other metals. Tin, either in alloy or pure form, has long been used in food preparation and storage. Bronze was used for cooking and storing vessels as well as for implements for eating. Pewter plates and dishes were on the tables of rich man and commoner for many hundreds of years. Food preparation and canning has depended in a major way on tin plate for more than a century. Thus, apart from the naturally occurring tin content of plants and animals, man has for many generations consumed food which has been in contact with, and very often absorbed, tin from culinary utensils. The metal has stood the test of time well. Only in some exceptional cases has its presence in food resulted in serious poisoning. Long-term effects due to low-level contamination are doubtful. There is even some evidence that tin may be an essential trace element in some animals, possibly including man.

CHEMICAL AND PHYSICAL PROPERTIES

The Romans knew tin by the *stannum* and it has been given the symbol Sn by chemists. Its atomic weight is 118.7 and atomic number 50. Its specific gravity is between 5.8 and 7.3, depending on the proportion of the three crystalline forms present. These forms are grey (cubic), white metallic (tetragonal) and white rhombic (tin brittle). It is a soft, white, lustrous metal which can be rolled easily into foil and extruded into tubes. Its melting-point is 231.9 °C. It is highly resistant to corrosion.

Oxidation states of tin are 2 and 4. It forms two series of compounds: the stannous and stannic. Among the more important inorganic compounds are the two chlorides, tin oxide, sodium chlorostannate, sodium penta-chlorostannate, stannous fluoride and sodium pentafluorostannate.

The metal also forms a large number of organic compounds, the organotins. The industrially more important of these include the alkyl tins as well as the phenyl compounds.

PRODUCTION

Though tin is present in traces in most soils, it is commercially produced in quantity in only a few places in the world. The principal ore is cassiterite or tinstone, SnO_2. Tin is also found in other ores in combination with other metals, such as tungsten. Extraction of tin from these mixed ores is difficult, but is undertaken when the quantity of the other metals present is sufficient to justify the high processing costs.

Today, tin is produced on a large scale in Malaysia (the world's leading producer), Bolivia, Thailand, Indonesia and the USSR. Smaller amounts are still produced in some of the ancient mines such as those of Cornwall and in several other countries including Zambia and Zaïre in Africa. Total world production was about 180 000 tonnes in 1973.

Tin is an expensive metal and its high cost has stimulated a considerable effort to recover and recycle the metal. Today, this 'secondary tin' meets some of the demands of industry for the metal.

USES

Upwards of 50 per cent of the world production of tin is used for plating. Tin plate is manufactured either by electrodeposition or by dipping and hot-wiping. In the latter, tin, either pure or containing other metals such as lead as a flux, is melted and then spread by wiping over steel, iron, copper and other surfaces which are to be plated. It is an operation easily carried out by experienced workers and was formerly standard practice in some large kitchens for restoring the worn surface of tinned cooking utensils. Very thin coatings of tin over other metals may be produced in this way. Such coatings, because of their corrosion resistance and the fact that they are easy to solder, are used extensively on food containers, forming an inert barrier between foodstuffs and the more easily attacked surfaces of the constructional metal. Processing equipment in the food and dairy industry, as well as a variety of catering equipment, are tin-plated for the same reason. Electrodeposited coatings are employed in the manufacture of engineering and electrical components and for many other industrial applications where corrosion resistance is important. Electroplated mild steel is used in enormous quantities in the manufacture of the 'tin' can. Alloys of tin with other metals, for example bronzes of various kinds, find use in a variety of industries. Pewter and the various related alloys of tin with antimony, copper and some other metals, such as Monel and Britannia Metal, are used largely for ornamental and some catering purposes.

About 5 per cent of the total tin produced is consumed in the chemical and related industries. It is employed in the manufacture of glasses and enamels, as a reducing agent in chemical processes, in dyeing and in textile printing. Some use has been made of inorganic tin compounds in the pharmaceutical industry, especially as an additive to toothpaste because of its possible cariostatic properties. Organic tin compounds are used as stabilisers in polyvinyl chloride plastics as well as in chlorinated rubber paints. Triphenyl tin and related compounds have been used as fungicides, insecticides and antihelminthics for farm animals.

TIN IN FOOD AND BEVERAGES

Normally the level of tin in food and beverages is low. When higher than normal levels are encountered these can be traced to contact with tin-lined utensils and containers or, in some cases, with PVC plastic wrappers. Most cereals, vegetables and meats contain about 1 mg/kg or less of tin, in the absence of external contamination. The distribution of the metal within a particular type of foodstuff may not be uniform. Wheat flour has been shown to contain 1.2 mg/kg and bran 1.6 mg/kg, while the outer pericarp contained 3.9 mg/kg.[15]

The average daily intake in the UK from a normal diet has been estimated to be 187 μg for an adult.[16] In a long-term study of a group of people apparently consuming a diet containing a higher level of external contamination from food canning and processing, higher levels of 1.5–3.8 mg of tin were found to be consumed each day.[17]

Increases in tin content in foods resulting from contact with the metal during processing and especially storage depend on a number of factors. As Monier-Williams[18] has pointed out, acid conditions increase solution of the metal from cans; corrosion is greatest at about pH 4.[19]

Storage at high temperatures and the presence of nitrates will also increase uptake of tin by food from the container. This matter has already been discussed in an earlier chapter dealing with the overall question of contamination of food by metal containers. In addition, it should be mentioned that tin contamination of food may also occur from the use of stannous salts as additives. The stannous ion has been found to prevent the release of other metals during storage in metal containers. It also maintains the level of ascorbic acid in the food.

Beverages have been shown on occasion to be contaminated with organotin compounds picked up from plastic containers. A level of less than 1 mg/kg has been reported in one case.[19]

Little tin occurs in drinking-water. It has been suggested that the use of bronze fittings could result in higher levels than the 1 μg/litre or less that has been reported. However, in the absence of industrial pollution or contamination of river water or well-water by sewage sludge used for agricultural purposes, there is little likelihood of appreciable levels of the metal occurring in potable water supplies.

METABOLISM AND BIOLOGICAL EFFECTS OF TIN

Rats have been shown to require tin at a rate of 1–2 mg/kg body weight each day in order to grow normally. There is no direct evidence that man has a similar need, but the universal distribution of the metal in human tissues and in food suggests that indeed tin may be an essential nutrient.

It is probably a tribute to the non-toxicity of tin and the confidence of man in its suitability for his culinary requirements that the metal has so long served him in the kitchen and food factory and that so few cases of food poisoning due to contamination by tin have been reported. For all practical purposes tin is a safe and efficient metal for the food-related purposes for which is it customarily used.

The absorption of tin from the diet is very limited. Animal and human experiments have shown that only about 1 per cent of tin in food is absorbed from the digestive tract.[20] There are differences in the levels of absorption of the different chemical forms of inorganic tin, with absorption of Sn(II) being about four times that of Sn(IV).[21]

Absorbed tin is rapidly excreted initially, but a small amount may have a biological half-time of about 100 days. Excretion of absorbed tin is mainly in urine with some via the bile.

Retained tin is distributed in kidney, liver and bone with little in soft tissues. The main deposit is found in the skeleton.

High levels of tin in ingested foods have been known to cause acute poisoning. The toxic dose for man has been reported to be 5–7 mg/kg of body weight.[22] Nausea, vomiting and other symptoms have occurred when levels of 250 mg/kg of tin in food have been consumed.[23]

However, in another investigation, levels of as high as 500 mg/litre of tin in fruit juice did not cause gastric upset. In fact, there can be considerable individual variation between persons in their reaction to different levels of ingested tin. Studies of volunteers who consumed about 200 mg/kg of tin in canned food over an extended period did not find evidence of a long-term effect.[24]

There is some evidence, however, that long-term effects of tin ingestion may occur in some animals. An experiment in which rats were given 5 mg/litre of tin in drinking-water from weaning to natural death showed no gross effects but apparently reduced longevity in females.[25] Another long-term experiment indicated the development of anaemia in rats fed on a tin-containing diet.[26]

The legally permitted maximum level of tin in several countries is 250 mg/kg. Concern has been expressed[22] that this limit does not allow an adequate safety margin since the standard 70 kg adult consuming 1 kg of such food would consume what is believed by some toxicologists to be an excessive amount of tin.

ANALYSIS OF FOODSTUFFS FOR TIN

The determination of tin in biological material presents a number of problems. None of the spectrophotometric methods are very specific or accurate at low concentrations of the metal. Colour development using dithiol, catechol violet or quercetin is used in a number of methods, but no particular reagent seems to be superior to any other.[27]

Today, atomic absorption spectrophotometry is widely used, but this technique also has disadvantages. Since tin oxide is not readily broken down in the flame, special systems have to be employed. Air–acetylene is not recommended because of the wide range of cationic and anionic interferences observed. Phosphoric and sulphuric acids, aluminium, copper and the alkali and alkaline earth metals are among the interfering substances recognised. In the absence of other elements, the air–hydrogen flame has been found to give high sensitivity. However, for mixed samples the nitrous oxide–acetylene flame is recommended, but sensitivity is low.[28] More recently tin has been analysed with success at low levels in foodstuffs by more advanced instrumental techniques, such as neutron activation[29] and X-ray fluorescence.[30]

9

Transition Metals

The transition metals are of considerable interest to biological and medical scientists as well as to the food manufacturer. The group includes many metals with important biological functions, such as iron, copper and chromium. These metals are also important industrially for the manufacture of processing and storage equipment and are often found as contaminants of food.

The transition elements are commonly defined as those which have partly filled d or f shells in any of their commonly occurring oxidation states.[1]

They are subdivided into three main groups: the main transition or d block elements, the lanthanide elements and the actinide elements. All have certain general properties in common, among which are the following:

1. They are metals, nearly all of which are hard and strong with high melting- and boiling-points, and they are capable of conducting heat and electricity well.
2. They form alloys with one another as well as with some other metals.
3. Almost all exhibit variable valency and their ions and compounds are coloured in one or more of their oxidation states.

The first two properties account for the usefulness of the transition metals to the manufacturer and food processor; the third is of significance especially from the point of view both of their biological activities as well as of our ability to detect and measure their concentrations in biological materials.

It is the main transition group or d block that is of particular interest to us here. This group contains those elements is which the d shells only are partially filled. Scandium, with an outer electron configuration of $4s^2 3d$, is the lightest member of this group. The eight succeeding elements—titanium, vanadium, chromium, manganese, iron, cobalt, nickel and copper—have partially filled $3d$ shells, either in the ground state of the free atom (all except copper) or in one of their more chemically important ions (all except scandium). This group of elements makes up the first transition series. They number elements 21 to 29 in the Periodic Table. The next—number 30, zinc—is sometimes considered to be a transition element, since it shares some of the properties of the preceding metals in the Table, but strictly speaking it is not so. Its outer electron configuration is $3d^{10}4s^2$, and it forms no compound in which the $3d$ shell is ionised. Thus it will be considered in a separate chapter of this study.

A second transition series begins with element number 39, yttrium. This has a ground state outer electron configuration of $5s^24d$. It is followed by eight elements—zirconium, niobium, molybdenum, technetium (an interesting element only recently discovered, though its existence has been suspected for a long time; it is recovered from waste fission products after removal of plutonium and uranium), rhenium (very similar to technetium and also isolated relatively recently), palladium and silver. Of this series, the metals of most consequence with regard to food are molybdenum and silver. These will be discussed in some detail later, but the others, which include some of the rare metals, will not be considered further here.

There is a third transition series, which begins with hafnium, which has the ground state outer electron configuration $6s^25d^2$ and includes the important (commercially at least) metals, platinum and gold. This series is not of particular interest from the point of view of food. Likewise the lanthanide and the actinide series, running from elements 58 to 71 and 90 to 103, are not of immediate interest here. However, it should be noted that the actinide series includes the radioactive elements thorium, uranium and plutonium, all of which are of extreme importance as environmental contaminants. The elements we will consider in some detail are those of the first transition series, apart from scandium, as well as molybdenum and silver in the second series. We will begin with copper, since it is one of the metals which has been longest in use and has always been associated with man's culinary activities.

COPPER

Copper was among the first metals to be used by man in a pure state, for not only is it easily extracted from its ores by primitive metallurgical techniques, but it is also found in elemental form on occasion. Its ores are, moreover, abundant and usually coloured and so are easily recognised even by the unskilled prospector. Because of its soft physical nature copper was probably used originally mainly for ornaments. It was not until the discovery of its ability to form alloys with other metals that it began to be employed on a large scale for making implements of various kinds, including weapons, as well as for vessels for culinary purposes. The Bronze Age was founded on the discovery and development of production of an alloy of copper with tin. In India and China the combination of copper with zinc gave the alloy which played, and still plays, a major role in the lives of millions of people. Ever since civilisation began, copper has been associated with food in one way or another.

CHEMICAL AND PHYSICAL PROPERTIES

Copper has an atomic weight of 63.54 and is number 29 in the Periodic Table of the elements. Its density is 8.96. It melts at $1083°C$. It is a tough though soft and ductile metal, second only to silver in its thermal and electrical conductivities. These are the properties that account for its usefulness in the electrical industry and for the manufacture of utensils of many kinds. Copper is resistant to corrosion in the sense that on exposure to air it is only superficially oxidised, sometimes giving a green coating of hydroxo-carbonate and hydroxo-sulphate, the well-known verdigris of copper-domed buildings.

Oxidation states of copper are normally 1 and 2 and it forms two series of compounds: Cu(I), cuprous and Cu(II), cupric. There is evidence that Cu(III) can also occur in crystalline compounds and complexes. The metal reacts with oxygen at red heat to give CuO and, at higher temperatures, Cu_2O. Copper forms a wide range of cuprous and cupric inorganic salts. The cupric compounds are the most important and most cuprous compounds are readily oxidised to the corresponding cupric form. These are usually water-soluble and coloured. Perhaps the commonest cupric salt is the hydrated sulphate, $CuSO_4.5H_2O$. The hydroxide, $Cu(OH)_2$, is obtained as a bulky blue precipitate on addition of alkali to cupric solutions. In ammonical solutions a deep blue tetramine complex is formed, $(Cu(NH_3)_4)^{2+}$. The ability of copper to form complexes with amines and other ligands is a distinctive characteristic and accounts for many of its biological activities. Such complex formation is used in several analytical procedures for the metal. Probably the best known of these is the blue solution produced by addition of tartrate to Cu^{2+} solutions (Fehling's solution).

PRODUCTION AND USES

Copper and its ores are widely distributed in nature. The principal ores are malachite, azurite, chalcopyrite, cuprite and bornite. The main ores are principally sulphides, but oxides and carbonates as well as arsenides and chlorides occur. Many copper ores also contain other metals such as zinc, cadmium and molybdenum. The metal is extracted by oxidative roasting and smelting, usually followed by electrodeposition from solutions.

World production of copper is about 6 million tonnes. Demand varies depending on international economic issues, and a decrease in price and production is normally associated with trade recession. In addition today a tendency, for reasons of economy mainly, to replace copper with aluminium in the electrical industry has reduced world requirements and resulted in the closure of many of the smaller and less productive copper mines.

The primary use of copper, accounting for as much as half of world production, is in the manufacture of electrical goods, electric cables and similar equipment where excellence of electrical conductivity is required. It is also used extensively for plumbing and heating. Food processing and catering equipment is no longer made from copper as frequently as formerly, though it is still the metal of choice for some processes. Copper sheets for roofing, covering boat hulls and for similar protective roles are still produced though, again, not as frequently as formerly. Copper is also a component of many alloys with other metals such as tin, zinc, silver and cadmium. Brass is still used for domestic purposes in India and other parts of the world, though, as in the West, aluminium and other metals tend to replace it. Bronzes of various types, such as the phosphor-bronzes which incorporate copper, are used for making specialist items in machinery and fittings.

Copper salts have some agricultural and pharmaceutical applications. One of the earliest and very important agricultural uses was as Bordeaux mixture, a fungicide used to prevent blight in grapes and potatoes. Copper-containing pesticides are available. A cupro-arsenate mixture is used as a preservative on wooden power and telephone poles in many parts of the world where fungal damage is common.

COPPER IN FOOD AND DAILY INTAKE

Copper is widely distributed in foodstuffs and the general opinion of nutritionists is that the human diet always contains sufficient of the metal for our needs. Generally concentrations in food are about 1 mg/kg. The best sources of the metal are meat, liver, kidney, heart and other forms of offal, as well as fish and green vegetables. Refined cereals and cows' milk are poorer sources, with usually less than 100 μg/kg of copper.

Daily intake has been estimated to be between 1 and 3 mg for the normal adult diet. This corresponds to between 15 and 45 μg/kg of body weight and meets the WHO estimates of requirements, which are about 30 μg/kg body weight for adults, 40 μg/kg body weight for older children and twice this amount, 80 μg/kg body weight, for infants.[2] Klevay[3] is of the opinion that many modern diets do not meet these requirements. He believes that food today contains less copper than it did formerly, possibly due to modern methods of food processing and production. As a consequence, the level of copper in the diet is falling. Klevay believes that this change is of significance with regard to the level of incidence of heart disease in modern technological society.

Our intake of copper can be influenced by the quality of water we consume. Levels of naturally occurring copper in drinking-water may vary considerably, depending on the nature of soil and rock through which the water travels. In addition, the use of copper plumbing systems may add a considerable amount of the metal to daily levels of consumption. The tendency today to replace metal pipes with plastic is of significance in this regard.

ABSORPTION AND STORAGE OF COPPER
BY THE HUMAN BODY

Because of its metabolic importance, the level of copper, like that of iron and possibly some other essential inorganic nutrients, is carefully regulated by the human body. Gastro-intestinal absorption is normally governed by the status of the metal in the body. Absorption occurs mainly in the stomach. Normally about 30 per cent of ingested copper is absorbed. On a low-copper diet this level of absorption has been shown to increase to between 50 and 65 per cent.[4] The percentage absorbed decreases when levels of copper ingested are increased.[5]

Copper uptake by the body is also related to the presence of other metals in the diet. A low molybdenum level in food may result in retention of copper, while excess molybdenum provokes a considerable increase in copper excretion. Such antagonism between molybdenum and copper has been recognised also in animals.[6] There is also, possibly, a relationship between copper uptake and the level of zinc in food, with an antagonism similar to that of copper and molybdenum.

Copper is preferentially stored in organs such as liver, heart, brain, kidney and muscle. The liver of newborn infants contains about 30 mg/kg of copper, but this falls to about 5 to 10 mg/kg within a year. The adult liver contains about the same concentrations of copper. The same pattern is seen in kidneys and other organs.[7]

Whole blood normally contains about 1 mg/litre of copper. However, this level can vary considerably, especially in women. During pregnancy and after the intake of oral contraceptives, copper can rise to 2 or more mg/litre in blood. The rise reflects an increase in ceruloplasmin. A corresponding increase is seen in levels of copper in hair. The normal level of approximately 30 mg/kg may increase to 70 mg/kg in the hair of women who begin to use oral contraceptives.[8]

Urinary excretion in adults on a dietary intake of 2 mg of copper per day has been shown to be between 11 and 48 μg/day. The main excretion route in animals as well as in man is via the bile in faeces.

The biological half-time of copper seems to be about 4 weeks in normal subjects. Retention is much longer in persons with Wilson's disease.[9]

BIOLOGICAL EFFECTS OF COPPER

Very little copper exists in free ionic state in body fluid or tissues but is nearly all complexed with proteins. The principal copper-containing protein of blood is ceruloplasmin. This is a blue protein of molecular weight 150 000 containing 8 Cu(I) and 8 Cu(II) ions. It contains about 3 per cent of the total copper of the body. Ceruloplasmin is manufactured in the liver and is involved in regulation of the levels of copper in the body. It also acts as a transport agent for the metal. In addition, it plays an important role in the oxidation of Fe(II) to Fe(III), an essential step in the transport of iron and the manufacture of haemoglobin in the body.

Copper functions in a number of important enzymes, many involved with oxidation reactions. It is found at the active centre of several organic catalysts, where its ability to undergo reversible reduction, like iron and some of the other transition metals, permits it to function in a variety of redox reactions. Enzymes with which it is involved include cytochrome oxidase at the terminal stage of the major pathway for cellular oxidations, and tyrosinase, responsible for metabolism of the amino acid tyrosine and the production of melanin, the dark pigment of skin and freckles. Other copper-containing enzymes of the human body are various amine oxidases, uricase and the superoxide dismutases. An important copper-containing enzyme of plant tissue is ascorbic acid oxidase, an enzyme responsible for loss of vitamin C in harvested fruit and vegetables. Obviously, deficiency of copper in the diet or failure to absorb sufficient from the gastro-intestinal trace could result in reduced activity of some enzymes. There is no clear evidence that direct copper deficiency ever occurs in adult humans, because of its adequate distribution in foodstuffs. However, copper deficiency has been reported in children fed on milk only during recovery from severe diarrhoea[10] and during parenteral nutrition.[11] One of the symptoms of the deficiency was anaemia. It is possible that in some cases a less than adequate intake may occur in the diet of certain sections of the population. In addition a hereditary disease, Menkes' 'steely hair' syndrome, is known in which there is an inability to absorb sufficient copper from the diet. In this state the hair has a characteristic appearance, and serious and widespread defects in cellular metabolism occur. Another hereditary disease related to copper is Wilson's disease. This results in excessive uptake and accumulation of copper by the body, with a build-up especially in liver and brain.

The diseases of humans due either to deficiency or excess of copper absorption are paralleled by well-recognised states of toxicity and deficiency in animals. There are copper-deficient areas of soil in various parts of the world where neither plants nor animals can survive without special supplementation. Copper-deficient animals have bone defects; hair colour is lacking and haemoglobin synthesis is impaired; cytochrome oxidase activity is low. In addition, defects in protein synthesis, affecting principally the elastin of arterial walls, occur. This protein is poorly cross-linked and the arteries are weak. Copper toxicity in animals on copper-rich soils and pastures is also known. Characteristic of copper poisoning is the disease known as 'swayback' in sheep. But it should also be mentioned that, as in the case of humans, copper is an essential element for animals and, provided it is supplied in appropriate amounts, copper supplementation of the diet, especially of pigs, has been shown to produce beneficial effects on growth and general state of health.

TOXIC EFFECTS OF COPPER IN FOOD AND BEVERAGES

Copper salts ingested in large amounts have been shown to cause toxic effects in humans, as well as in animals. The effects are usually reversible.

Evidence for long-term effects is not convincing. Industrial exposure in the case of vineyard sprayers who accidentally inhale large quantities of copper sulphate when using Bordeaux mixture on vines has resulted in what is called 'vineyard sprayer's lung'. This resembles silicosis histologically and in some cases lung cancers have been found in association with copper deposits. However, it is not clear whether there was a strict causative connection between the metal and the cancer.[12] Suicide attempts in which upwards of 100 g of copper, usually as copper sulphate, were ingested resulted in jaundice and renal damage, but sometimes only in gastro-intestinal disturbances. However, though some fatalities have been reported,[13] copper sulphate is a powerful emetic and it is actually difficult to absorb sufficient of the metal to produce serious effects. In children about 100 mg of copper induces vomiting while adults may require 400 mg (1 g of copper sulphate).[14] Small, non-emetic additional intakes of copper often result from the ingestion of copper-contaminated food and drink. Prolonged retention and reheating of coffee in a copper boiler has been known to result in gastro-intestinal upset. Drinking water from the hot-water tap where copper plumbing and heaters are used may result in more than desirable intake of the metal.[15]

There have been reports in recent years of excessive uptake resulting from copper piping and connections in dialysis machines used for treatment of persons with renal disease. This has led to an increase of copper in the plasma with consequent haemolytic anaemia in one case. Fatalities have resulted from this cause.[16]

EFFECTS OF COPPER ON FOOD QUALITY

From the point of view of the food manufacturer, the presence of small amounts of copper in his products may be much more significant because of its effect on food quality than as a possible toxin. In the case of edible oils and foods containing unsaturated fats, copper ions can act as a catalyst in oxidation leading to rancidity, colour changes and other reactions. These, while highly undesirable from the point of view of appearance and acceptability, do not result in toxicity or a decrease in nutritive value of food.

METHODS OF ANALYSIS OF COPPER IN FOOD

The chemistry of copper has made it relatively easy to detect in foods, and colorimetric procedures for qualitative and quantitative analysis have been known for many years. The most common of these uses the yellow complex formed between copper and diethyl dithiocarbamate. The metal must first be extracted from the food sample with an organic solvent, usually amyl alcohol. Other colorimetric methods are used and most are reliable, and results obtained using them agree with those from modern instrumental techniques.

The favoured technique today is usually atomic absorption spectrophotometry. An acetylene–air flame is suitable and little interference by other elements is experienced. The Analytical Methods Subcommittee of the Society of Analytical Chemistry[17] has recommended that ready-to-drink beverages can be aspirated directly without destruction of organic matter, provided that the sample has been suitably diluted with water, to reduce soluble solids below 3 per cent. Semi-solid foodstuffs should be extracted with organic solvents and the extract aspirated directly. Solid foods may be wet digested or dry ashed and standard procedures for analysis followed. The use of flameless atomic absorption permits the detection of nanogram quantities of copper. This method has been particularly useful in the analysis of microlitre samples of biological fluids.[18]

Emission spectrophotometry, neutron activation and electrochemical methods such as pulse polarography and anodic stripping voltametry have all been used with success for the determination of copper in foods and other biological materials.

IRON

Iron is the second most abundant metal, after aluminium, and the fourth most abundant element in the earth's crust. The earth's core is believed to consist of iron and nickel and, to judge from the existence of many iron meteorites, iron is also abundant elsewhere in the solar system.

The importance of this metal in the physical world is matched by its importance to human life. Civilization, as we know it, would not exist without iron. We use more of it, in pure form and in alloys, than of any other metal. An era of man's development has been named the Iron Age, but in fact, our dependence on that metal has increased rather than decreased since the historians tell us that period ended.

The human body and indeed most living organisms have, during the course of evolution, come to depend on iron as a linchpin of existence. The production and utilisation of energy from the food we consume and the air we breathe are dependent on the presence of iron in cells and tissues. Without iron the blood pigments cannot be formed, nor will hundreds of other cellular reactions take place. It is interesting that of all the organisms that inhabit this earth, apparently only one group, the anaerobic lactic acid bacteria, possess no iron-containing enzymes and appear to be totally devoid of iron as well as of copper.

CHEMICAL AND PHYSICAL PROPERTIES

Iron, which has the chemical symbol Fe after its Latin name *ferrum*, is element number 26 in the Periodic Table, with an atomic weight of 55.85. It is a heavy metal with a density of 7.86. Pure iron is a white, lustrous metal which melts at 1528 °C. It is not particularly hard and is ductile and malleable. It is a good conductor of heat and electricity.

Chemically iron is quite reactive. In moist air it rapidly oxidises to give a hydrous oxide which is generally recognised as rust. Unfortunately, a skin of rust affords no protection to the underlying metal for the oxidised layer easily flakes off, exposing fresh metal surfaces.

Iron commonly forms two series of compounds: the ferrous, Fe(II), and ferric, Fe(III). Other oxidation states also occur, but are not of major biological importance. Three oxides of iron occur, FeO, Fe_2O_3 and Fe_3O_4, representing the Fe(II) and the Fe(III) as well as the mixed Fe(II)-Fe(III) oxide which occurs in nature as the black mineral magnetite. The metal dissolves in dilute mineral acids to form salts. With non-oxidising acids and in the absence of air, Fe(II) salts are obtained.

Of nutritional significance is the fact that many salts of iron, as well as the hydroxides, are relatively insoluble. Iron forms a number of complexes and this property accounts for many of its biological activities. The haem and the non-haem complexes, such as are found in haemoglobin and the ferredoxins, are of considerable importance in biological oxidation and in nitrogen and carbon dioxide fixation.

PRODUCTION

Though many soils contain as much as 4 per cent iron, the metal is not extracted from this source but from a number of widely distributed ores. The major iron-ores are haematite, Fe_2O_3, magnetite, Fe_3O_4, limonite, FeO(OH) and siderite, $FeCO_3$. The production of iron is a well-developed and highly skilled process, involving reduction by smelting with various forms of carbon.

USES

The purposes for which iron is used are so well known and so numerous that there is no need here to list them. Whether in the form of cast iron or as a component of steel, iron plays a major part in the construction of processing equipment, containers and various utensils used for food. It is through such equipment that most food contamination by iron comes about.

IRON IN FOOD AND BEVERAGES

Almost all foodstuffs contain some iron, though in certain cases the total amount present or its availability for absorption may not be sufficient for nutritional needs. There can be considerable variation between levels of the metal in different types of food. The range of iron content in some foods analysed in the UK is as follows (expressed as mg/kg:[19] fresh meat, 3.0–4.0; liver, 6.0–14.0; fish, 0.5–1.0; eggs, 2.0–3.0; white flour, 0.7–1.5; wholemeal flour, 3.0–7.0; oatmeal, 3.8–5.1; green leafy vegetables, 0.4–18.0; potatoes and other root vegetables, 0.3–2.0; fruit, 0.2–4.0; cows' milk, 0.1–0.4. It should be stressed that these figures represent total iron content of food. As will be pointed out later, actual availability for gastro-intestinal absorption will often be quite different.[20]

Canned fruits, vegetables and other foods may contain elevated levels of iron due to solution of the metal from the container walls. The pH of the can contents as well as the standard of canning and other factors will determine the level of metal taken up. Of particular importance with regard to iron uptake from cans is the level of nitrate in the contents. Some fruits such as the tropical pawpaw (papaya) may have such high levels of $NO_3{}^-$ that they cannot be canned successfully because of the excessive degree of iron solution that results. Some temperate climate fruits also present problems. In one study,[21] for instance, while canned grapefruit were found to contain 5.3 mg/kg (wet weight, solid portions) and pears 5.0 mg/kg, blackcurrants had 1300 mg/kg and pineapple 31 mg/kg of the metal. In the same study canned green beans were found to have 2.8 mg/kg, processed peas 9.9 mg/kg and mushrooms 5.1 mg/kg of iron.[21]

The problem of take-up of metal by canned foodstuffs and the shelf-life of such products has already been discussed. It is sufficient to note here that there can be considerable variations between the levels of metals in canned foods, depending on the degree of quality control exercised by the manufacturer. Generally, however, canned foods will contain more iron than fresh or frozen packaged commodities and a diet in which canned foods figure largely may lead to a relatively high intake of iron, as well as of some of the other packaging metals.

Whether or not the iron ingested in food will actually be absorbed and result in a high level of body iron is another matter, involving the question of availability of metals, and will be discussed below.

Iron may be present also in some foods as a result of deliberate fortification to compensate for losses during processing. While, for example, whole wheat grains contain about 40 mg/kg of iron, more than 75 per cent is lost in milling, giving a white flour containing about 10 mg/kg of the metal.[22] For this reason, several countries, including the UK and Sweden, require flour millers to 'fortify' their products with iron. The iron used is either in the form of ferric ammonium citrate or as metallic 'reduced iron'. There is considerable debate as to advisability of such fortification.

Iron also enters the diet in water. The metal is normally present in domestic water supplies, both from natural sources and as a result of the use of iron in water treatment and supply equipment. The average level of iron in the domestic supply in one American city was reported to be 166.5 μg/litre[23] and similar levels have been reported from urban supplies in other parts of the world. These levels are within the limit established by the US Public Health Service Drinking Water Standards of 0.3 mg/litre of iron in public water supplies. Far higher levels are often found where either the plumbing system is old or the water is allowed to stand in iron tanks or flow through cast-iron or galvanised service lines. Well and river water will generally show peculiarities of iron and other metal levels, depending on the type of bedrock and on whether pollution occurs or not.

It should be noted here that the presence of iron at higher than average levels in water and other beverages and in foodstuffs, apart from its implications for nutrition, is of importance from the point of view of aesthetics and food quality. Iron in more than trace amounts is an objectional constituent in water, either for industrial or domestic uses. Staining of fabrics by iron hydroxides and other salts can occur where iron-rich laundry water is used. In addition, the element gives a bitter or astringent taste to water and this can affect beverages made from this water.[22] Because of the ability of iron to form coloured complexes with various organic compounds, including naturally occurring tannins, the presence of iron in canned fruits, beverages and other foodstuffs can result in discoloration and render the product unacceptable to the public. In addition, iron, like copper and some other metals, can act as a catalyst for oxidative reactions of unsaturated carboxylic acids, leading to rancidity in fats and fat-containing foodstuffs. For these reasons both domestic users and food processors prefer iron-free water. If this is not available it may be necessary to use chelating agents to sequester the iron atoms. This can be a costly process but is essential where the presence of iron and other metals with similar chemical and physical properties could bring about spoilage of food products.

ABSORPTION OF IRON FROM FOOD

The mean daily intake of iron in the diet in the UK has been shown to be approximately 13 mg. Intake in the USA and other technologically developed countries is about the same, with at least 12 mg/day.

The human body has the ability to control to a remarkable extent its overall level of iron. Absorption as well as excretion are regulated by biological mechanisms, the details of which are not yet fully understood. There are about 4 g of iron in the average adult body. This total remains almost constant in health. In illness or in particular stages of life, such as during the reproductive period of a woman, reductions may occur and iron deficiency develop. Constancy is normally maintained by a careful conservation of the iron already absorbed, with urinary and other losses reduced to a minimum.

Absorption of iron from ingested food is related largely to the iron status of the body. It occurs mainly in the jejunum, but there is also a small amount of absorption in the stomach. Inorganic iron in food is usually present as ferric hydroxide complexed with various organic compounds such as proteins, amino acids and carboxylic acids. The complexes are first split, normally by digestive enzymes aided by the acid conditions of the stomach. The insoluble Fe(III) so released is then reduced to the more soluble Fe(II) form. This reduction is brought about by naturally occurring reducing substances which include ascorbic acid (vitamin C) and various sulphydryl compounds, such as the amino acid cysteine. The ferrous iron is now taken through the brush border of the intestinal mucosa. Once in the mucosal cells the iron may be either bound to a protein, a globulin (transferrin) and taken on further into the bloodstream, or it may be stored in the mucosa as ferritin. This is formed when iron combines with another protein known as apoferritin. It was formerly believed that absorption of iron from the gut was controlled by the level of this stored iron, in the mucosal cells. This theory of control was known as the 'ferritin curtain' or the 'mucosal block'. While, indeed, the level of iron stored in the body is clearly related to uptake, this rather simple mechanism is no longer believed to be at least the whole explanation. It would appear that a more complicated system operates, involving regulation of active transport through the mucosal cell wall. This system is in turn related to the status of body stores as well as to the activity of bone marrow, where iron is used to make haemoglobin. We know, for instance, that when body stores of iron are low, as in patients with iron deficiency anaemia, or when demands for additional iron are greatest, as in growing children and pregnant women, restrictions on mucosal absorption are lifted and iron absorption increases.

Various factors in the diet can affect iron absorption, as has been mentioned. These are related to the different steps of absorption mentioned above. Thus, for instance, anything which interferes with the splitting of the iron complexes in food or with the reduction of iron to the soluble Fe(II) state will limit absorption. The presence in food of substances which form insoluble complexes with iron are among such factors. Phosphate in cereals and in egg yolk, as well as oxalic acid in various vegetables, for example, can do this. Phytic acid (inositol hexaphosphate), which is present in cereals and especially in wholemeal flour and products made from it, can form strong, insoluble complexes with iron and other metals and seriously hinder uptake. On the other hand, some substances, also present naturally in foods, can help absorption by aiding reduction and increasing solubility. Ascorbic acid and certain amino acids have already been referred to in this regard. Ascorbic acid forms a soluble complex with iron and assists passage through the cell walls of the mucosa, especially under acid conditions. It is possibly the presence of cysteine in meat that accounts for the observation that absorption of iron from certain foods is increased if meat is eaten simultaneously with them.

The naturally occurring form of iron in meat and some other foods is haem, the metallo-porphyrin complex found in the blood and muscle respiratory pigments haemoglobin and myoglobin. Haem does not have to be split before it is absorbed into the mucosal cells, nor is it affected by phosphate nor phytates. The complex is split only after absorption and then the iron atoms enter the normal pathway of transport and utilisation.

Various body states and conditions can also affect levels of iron absorption by the intestine. 'Intestinal hurry' and various states of malabsorption can restrict entry. Actual levels of absorption of iron, consequently, will vary between individuals, and even in individuals at different times of life, depending on nutritional status and other factors, such as the overall composition of the diet. A WHO report[20] shows that levels of absorption from different foodstuffs may vary from less than 1 per cent to over 20 per cent. There is a gradation of efficiency of absorption of iron from foods of vegetable origin to those coming from animals, with rice at the lower end of the scale and veal muscle at the top. Absorption of iron from a mixed Western diet has been estimated to be about 6 per cent in normal males, 14 per cent in normal females and upwards of 20 per cent in iron-deficient subjects.

Once iron has been absorbed, as has been said, it may be stored or utilised for various body functions. For utilisation, the iron must be transported away from the mucosal cells. This step involves oxidation of the iron once more from the ferrous to the ferric state. The iron is then bound to a transport protein known as transferrin. This step is assisted by the copper protein ceruloplasmin. The iron travels as an Fe(III)–transferrin complex to the spleen and liver where it may be stored once more as ferritin. Some of the iron will go on to the bone marrow, where it is utilised in the formation of haemoglobin. Some 70 per cent of the total iron in the body is located in the erythrocytes of the blood. Another fraction is concentrated in the oxidative enzymes of tissue cells. As has been mentioned earlier, the presence of lead in the body can interfere with the formation of haemoglobin at a number of stages of the complex synthetic processes involved. Lead can block the incorporation if iron into the porphyrin molecule and also interfere with synthesis of the porphyrin molecule itself.

EFFECTS OF IRON IN FOOD ON BODY METABOLISM

Iron is an essential nutrient. An intake of 10 mg/day for men and 12 mg/day for women is recommended by the UK Department of Health and Social Security. In certain conditions, as during pregnancy and at active stages of growth, intake requirements are larger. Though, as has been noted, intake of iron by the body is generally carefully regulated by the actual body status of the metal, sometimes an excessive intake may occur. The resulting increase in body stores gives rise to a condition known as siderosis. In this stage ferritin molecules, together with extra iron, become conglomerated and form haemosiderin. This can result in accumulation of iron in the liver to well over the normal level of 1 g/kg. The haemosiderin can be readily detected in tissue sections by adding ferricyanide, when a distinctive blue colour (Prussian blue) is formed. Such iron-overload siderosis is not uncommon among Africans in the southern half of the continent. It has been suggested that this is related to the consumption of large amounts of iron-contaminated home-brewed beer. The beer is commonly brewed in discarded oil drums and levels of 300 mg/litre of iron have been detected in such beverages. Daily consumption of 4 litres or more of beer is not uncommon and autopsies have revealed high iron levels in livers of habitual drinkers.[24] A study in another part of Africa has reported similar results.[25] A similar condition has been observed in Normandy, where cider may contain upwards of 16 mg/litre of iron. It has been suggested that the presence of iron in alcoholic beverages may be linked to the development of cirrhosis of the liver.

The use of cast-iron pots for preparing food has also been suggested as a cause of haemosiderosis among some Africans and other people.[24] A study in Papua–New Guinea[26] showed the extent of iron enrichment of food which can come from cooking utensils. In general the extra iron contributed to the diet in this manner is considered to be an advantage to health and may be responsible for the low level of iron-deficiency anaemia encountered in some people who have not abandoned traditional cooking utensils for teflon-coated saucepans or similar modern utensils.

ANALYSIS OF FOOD FOR IRON

The analysis of iron in biological materials is easily performed by colorimetric as well as by spectrophotometric and other instrumental methods. The ability of transition metals to form coloured complexes has been utilised in a number of different colorimetric methods. Low levels of iron can be determined without difficulty by either flame or flameless atomic absorption techniques. Normally an air–acetylene flame is found to be effective, with no interference from other inorganic species. Food samples may be prepared for analysis by wet digestion in acid solution or by dry ashing and subsequent solution in dilute acid. However, where direct analysis of liquid foodstuffs is carried out, interference due to viscosity and surface tension (in vegetable oil) or from dissolved carbon dioxide (beer) may cause problems. The use of a standard addition technique will help to overcome these difficulties. Degassing in the case of carbonated beverages is helpful.

Interference from citric acid has been reported to suppress the absorbance by upwards of 50 per cent when the acid is present at 200 mg/litre concentration. Adjustment of burner height and addition of phosphoric acid helps to minimise this source of interference. The use of a nitrous oxide–acetylene flame has been found to remove most interference.

CHROMIUM

Chromium is widely distributed in the earth's crust, forming about 0.04 per cent of the solid matter. Individual rocks and soils vary greatly in their chromium content but there seems to be enough in all soils to meet the nutritional requirements of plants and animals, including man. Industrially, chromium is a 'new' metal, finding its place in modern steels and other alloys and in plating. It has many important industrial uses and it is these, especially in the manufacture of stainless steel and plated food-processing equipment and culinary utensils, that contribute a great deal to dietary intake of the metal. Chromium is now recognised as an essential nutrient for animals, including man. There is a growing interest among medical scientists in its metabolic roles and in the implication both of excess and of deficiency of the metal.

CHEMICAL AND PHYSICAL PROPERTIES

Chromium has an atomic weight of 52 and is number 24 in the Periodic Table of the elements. It is a heavy metal with a density of 7.2 and a melting-point of approximately $1860\,^{\circ}C$, the exact temperature depending on the crystal structure of the sample in question. It is a very hard, white, lustrous and brittle metal, and is extremely resistant to many corrosive agents. Of its nine possible oxidation states, three are of practical significance, namely 2, 3, and 6. The most stable and important oxidation state is Cr(III), which gives a series of chromic compounds, such as the oxide, Cr_2O_3, chloride, $CrCl_3$ and the sulphate, $Cr_2(SO_4)_3$. Industrially important Cr(III) salts are the acetate, citrate and chloride. Cr(III) also forms large numbers of relatively kinetically inert complexes. An interesting example, widely used in preparative chemistry to precipitate large cations, is Reinecke's salt, $NH_4[Cr(NCS)_4(NH_3)_4(NH_3)_2].H_2O$.

In higher oxidation states almost all the chromium compounds are oxo-forms and are potent oxidising agents. The Cr(VI) compounds include the chromate, CrO_4^{2-}, and the dichromate, $Cr_2O_7^{2-}$, ions. Important compounds of this type include lead, zinc, calcium and barium chromate, as well as sodium and potassium chromate and dichromate.

The relatively unstable Cr(II) chromous ion is rapidly oxidised to the Cr(III) form. Few of the other forms of chromium, such as Cr(IV) or Cr(V), are known and they do not appear to be of biological significance.

PRODUCTION AND USES

Though chromium is present in most soils and rocks, its only important commercial ore is chromite, $FeOCr_2O_3$, a mixed ore which may contain upwards of 55 per cent chromic oxide. Reduction of the ore with carbon produces ferrochrome, a carbon-containing alloy of chromium and iron. Pure chromium is produced from this, either by electrolytic treatment or by conversion into sodium dichromate by treatment with hot alkali and oxygen and subsequent reduction to the pure metal by reaction with aluminium.

The major application of chromium is in the metallurgical industry as an alloying element in the production of stainless steel and in the plating of metal objects to protect against corrosion. Both ferrochrome and chromium itself are employed in the alloy industry. Chromium is also used for the production of pigments and in the printing industry. Chrome tanning is a traditional way of preparing leather. Chromate compounds are used as water additives to prevent corrosion and it is probable that their presence in cooling water accounts for a significant amount of industrial chromate emission to the atmosphere.

CHROMIUM IN FOOD AND BEVERAGES

Chromium is found, at least at low levels, in most beverages and foodstuffs. Amounts present range from trace to about 0.5 mg/kg. The average daily dietary intake has been estimated to be between 50 and 80 μg.[27] Levels in institutional diets in the USA were found to range from 0.175 to 0.470 mg/kg.[28]

The concentrations of chromium in a number of different classes of foods are as follows (in mg/kg): seafoods, 0–0.44; cereals, 0–0.52; fruits, 0–0.2; vegetables, 0–0.36; meats, 0.02–0.56.[29] Milk contains 0.01 and butter 0.17 mg/kg.[30]

Refining of foodstuffs can result in considerable loss of chromium. Raw sugar, for instance, contains 0.3 mg/kg, while white refined sugar has only 0.02 mg/kg. Similarly, unpolished rice contains 0.16 and polished rice 0.04 mg/kg. The difference found between whole wheat and white flour is less significant, from 0.05 to 0.03 mg/kg.[30]

The use of stainless steel cooking utensils has been shown to result in some cases of increased chromium in foodstuffs, to a maximum of 3.5 mg/kg.[31] Cigarette smoking can contribute to the daily intake of the metal. An average of 1.4 μg of chromium per cigarette has been reported.[31]

Levels of intake of chromium from domestic water supplies depends on the source of the water and the nature of treatment and reticulation systems. River water has been reported to contain between 1 and 10 μg/litre.[32] Urban supplies can be higher, with up to 80 μg/litre in some cases.[33] However, the average level for community water supplies in the USA have been estimated to be 2.3 μg/litre.[34] This is considerably below the US Public Health Service Drinking Water Standards (1962) limit of 50 μg/litre.

A potential source of increased concentrations of chromium in food, as of many of the other heavy, industrially important metals, is sewage sludge. Levels of as high as 8 g/kg have been reported in dry sludge.[35] The use of such material for agricultural purposes could result in higher than average uptake of chromium by plants as well as by animals.

ABSORPTION AND STORAGE OF CHROMIUM BY THE HUMAN BODY

Gastro-intestinal absorption of chromium is related to the chemical form of the element in food. Animal experiments indicate that trivalent chromium is poorly absorbed, at about 1 per cent of the total amount ingested, while chromates, at 2 per cent, are a little more readily absorbed.[36] Organically bound chromium, such as in the 'glucose tolerance factor' (GTF), which will be discussed later, is probably more readily absorbed from the gastro-intestinal tract. Studies in which urinary excretion were compared with daily dietary intake indicate an overall level of absorption of about 10 per cent of chromium in human food.[37]

Once absorbed, chromium is rapidly cleared from the blood. Much is distributed to various organs, in particular the liver, where it is found in trivalent form. The chromium concentration in all tissues decreases with age, except in the case of the lungs where environmental contamination may account for continuing higher than average levels. Excretion of absorbed chromium occurs mainly in urine. It is of interest that the highest concentration in humans has been reported to be in hair, with values from 0.2 to 2.0 mg/kg.[38]

BIOLOGICAL EFFECTS OF CHROMIUM

Chromium is one of the essential trace metal nutrients of animals and man.[39] Its principal role appears to be in helping to maintain normal glucose tolerance in the body. Deficiencies of the metal in the diet are associated with altered glucose and lipid metabolism and may result in diabetic and arteriosclerotic diseases.

A chromium-containing fraction has been prepared from yeast and has been recognised as a glucose tolerance factor. The exact structure of this factor is not known, but it is probably a complex of trivalent chromium, nicotinic acid and various amino acids.[40] In the absence of this factor, it has been shown that glucose injected into the bloodstream is removed much slower than normally.[41] This is the same kind of response observed in the absence of insulin. The chromium in the glucose tolerance factor is thought to react with the insulin hormone in some way to potentiate its effect, possibly by helping to bind it to cell membrane receptors. The importance of chromium in carbohydrate metabolism and other metabolic activities has only recently been recognised but it is now the subject of intense investigation. While chromium deficiency has been described in man and some of its effects recognised, so far daily human requirements have not been established.[42]

A good deal is known about the acute and chronic effect of industrial exposure to chromium and its compounds. Chronic ulcers are induced by the corrosive action of hexavalent chromium and chromic compounds in tannery workers. Allergic eczema and other forms of dermatitis, as well as cancer of the nose and lungs, occur among workers with chromium and its compounds.

TOXIC EFFECTS OF CHROMIUM IN FOOD AND BEVERAGES

There is no evidence that the chromium normally present in the diet produces adverse effects on health, whether the metal is present in the original foodstuffs or comes from the use of stainless steel utensils for cooking.[43] However, ingestion of large amounts of potassium dichromate has resulted in death. Smaller doses have been said to cause kidney and liver damage.[44]

METHODS OF ANALYSIS OF CHROMIUM IN FOOD

Colorimetric methods are available for the estimation of chromium in food. Problems arise due to interference by other ions. Chromium in urine or in organic extracts of food can be estimated using 1,5-diphenylcarbazide.[45] Flame emission photometry using an acetylene–nitrous oxide flame can also be used but is generally less sensitive than the diphenylcarbazide method.

Atomic absorption spectrophotometry, with either an air–acetylene or a nitrous oxide–acetylene flame, is widely used. However, various metals such as cobalt, iron and nickel, especially in the presence of perchloric acid, have been found to cause depression of absorption. Interference by copper, barium, magnesium and calcium in the air–acetylene flame is also reported to occur.[46] Careful selection of flame conditions can overcome some of this interference.

MANGANESE

Manganese is element number 25 in the Periodic Table and has an atomic weight of 54.94. The chemical and physical properties of manganese are very similar to those of iron, which it immediately precedes in the first transition series. However, it is harder and more brittle than iron and is also less refractory, with a melting-point of 1247 °C. It is a reactive metal which dissolves readily in dilute, non-oxidising acids. It burns in chlorine to give $MnCl_2$ and reacts with oxygen at high temperatures, producing Mn_3O_4. It also combines directly with boron, carbon, sulphur, silicon and phosphorus.

Manganese has several oxidation states, some of which, however, are of no practical consequence since they occur so rarely. The most stable and important is the divalent state, Mn(II). It forms a series of manganous salts with all the common anions. Most are soluble in water and crystallise as hydrates. Mn(II) also forms a series of complexes with chelating agents such as EDTA, oxalate and ethylenediamine.

The chemistry of Mn(III) is not extensive. In aqueous solution it is unstable and is readily reduced to Mn(II). Similarly, Mn(IV) compounds are of minor significance. However, the oxide MnO_2, pyrolusite, occurs in nature and is one of the chief manganese ores.

Mn(VI) is found only in the important manganate ion, $(MnO_2)^{2-}$. Similarly, Mn(VII) is best known in the permanganate ion, $(MnO_4)^-$. Potassium permanganate is a common and widely used compound of this group. It is a powerful oxidising agent, and has extensive pharmaceutical as well as chemical applications. Because of its wine-like colour and its availability, potassium permanganate in the form of Condy's fluid has been involved on a number of occasions in poisoning incidents.

PRODUCTION AND USES

Manganese is relatively abundant, making up about 0.09 per cent of the lithosphere, and is the twelfth most common element. Of the heavy metals, iron is the only one which is more abundant than manganese. Ores of manganese occur in a number of substantial deposits, the most important of which is pyrolusite, the Mn(IV) oxide. It is mined extensively in parts of Africa, the USSR and Canada. Production from oxide and other ores (principally carbonate) is by roasting in air followed by reduction with aluminium.

Manganese has three principal industrial uses—in steel making, the manufacture of electrical accumulators and as an oxidising agent in the chemical industry.

MANGANESE IN FOOD AND BEVERAGES

Manganese is widely distributed in plant and animal tissues and occurs in all food and water supplies. The manganese content of most foods does not normally appear to be subject to wide variation, but considerable differences in content of manganese in the different kinds of foods exist. This appears clearly in a detailed study of manganese in the diet recently completed in the UK.[47] While for example cereals, especially if unrefined, are rich in the metal, with a range of 2.4–14.0 mg/kg, meat and fish are relatively poor (0.35–1.1 and 0.45–1.3 mg/kg respectively). Vegetables, especially root vegetables (0.5–2.2 mg/kg) are somewhat richer than animal-based foodstuffs. Fruit may have as much as 3.6 mg/kg. An interesting, and in the UK context very important and rich dietary source of manganese, is tea. Dry tea contains between 350 and 900 mg/kg and beverages in the diet (including tea, coffee and a variety of soft drinks) have a range of 7.1–38.0 mg/kg of manganese. An analysis of the 'total diet' gave the following estimate of intake per day for the average adult:

	mg/day
Cereals	1.56
Meat	0.09
Fish	0.02
Milk	<0.04
Fats	<0.02
Vegetables	0.41
Fruit and Sugars	0.28
Beverages	2.24
Total	4.6

The UK average daily intake of 4.6 mg is within the range of 2–9 mg which has been calculated for other countries.[48] However, actual levels of intake will depend on the dietary pattern. Where tea does not figure largely in the diet, cereals will normally be the major source of manganese intake. Even then levels of intake will depend on whether the cereals are highly refined or not. Processing causes a considerable reduction of levels of manganese, not only in cereals but also in other foods, such as sugar. Raw sugar, for example, contains 0.4 mg/kg and white refined sugar less than 0.05 mg/kg of the metal. Losses with cereals are somewhat less, with 4.7 mg/kg in corn and 2.05 mg/kg in cornflour, 2.8 mg/kg in whole rice and 1.53 mg/kg in the polished product.[49]

BIOLOGICAL EFFECTS OF MANGANESE

Manganese, like many of the other heavy metals, does not appear to be very efficiently absorbed by the body. Possibly about 10 per cent is the maximum transferred through the gastro-intestinal wall to the bloodstream from ingested food, but actual levels of absorption will depend on the total composition of the diet and probably on the chemical form of the metal in the food. Absorbed manganese is efficiently excreted through the bile and appears in the faeces.

The adult body contains about 8 mg of manganese in total, with the highest concentrations in muscle and liver.[50] The lungs may also have relatively high concentrations, depending on the level of industrial exposure to metallic dusts. Overall, human tissue contains less than 1 mg/kg on a dry weight basis.

Manganese is an essential element for the maintenance of normal health in animals and man. In its absence a number of well-defined deficiency symptoms have been observed in animals. These are particularly related to skeletal abnormalities. In these cases the organic matrix of bones and cartilage tends to develop poorly, with a decrease in the normal levels of various essential components of cartilage, such as hexuronic acids and chondroitin sulphate.[51]

Manganese plays an important role in cellular metabolism. A number of enzymes specifically require manganese for their function. These include enzymes for the synthesis of the mucopolysaccharides, which could account for deficiency symptoms related to cartilage defects. In addition, an enzyme involved in energy metabolism, pyruvate carboxylase, contains four tightly bound manganese atoms for each molecule of the vitamin biotin present. Manganese is a component of one of the recently much studied superoxide dismutase enzymes. These appear to play a role in protecting the organism against the deleterious effects of the superoxide radicals and may afford protection against certain forms of cancer.

Several other enzymes require manganese. These include arginase, some phosphotransferases and probably certain nucleases and DNA polymerases.

TOXIC EFFECTS OF MANGANESE ON HUMAN HEALTH

While adverse effects on health produced by manganese in miners and other workers are well known, there is no clear evidence that manganese in food is toxic. An acute 'metal fume fever' in manganese miners affects the central nervous system and may lead to 'manganese madness'. However, no such neurological effects have been reported from oral ingestion of the metal by man or animals.[52] In fact, manganese appears to be one of the least toxic of metals.

ANALYSIS OF FOOD FOR MANGANESE

Like the other transition metals, manganese forms a number of complexes with organic reagents, which are used in colorimetric determination of the metal.

Atomic absorption spectrophotometry is a widely used analytical method for the metal. Sample preparation may be by extraction from foodstuffs by organic solvents or by wet digestion or dry ashing. Interference from several other metals has been reported, but this seems to depend on the solution conditions. In the recommended nitrous oxide–acetylene flame, interference can be suppressed by adding an excess of a refractory element such as aluminium.[53]

COBALT

Cobalt is one of the most fascinating of the essential trace metals, with several unusual facets. It is a relatively rare element, making up only about 0.001 per cent of the lithosphere. It occurs in animals and man in minute amounts, and the daily intake by the human body is not much more on average than 0.1 μg. It is an integral part of vitamin B_{12}, of which the body has not much more than 5 mg, but if this vitamin is deficient, pernicious anaemia, a usually fatal disease in the absence of treatment, results. Its role in vitamin B_{12} appears to be the sole function of cobalt in the human body. Vitamin B_{12} has the distinction of being formed principally in microorganisms. Plants contain little or none. Thus a hazard for strict vegetarians is the possibility of an inadequate intake of vitamin B_{12}.

CHEMICAL AND PHYSICAL PROPERTIES

Cobalt has an atomic weight of 58.9, and an atomic number of 27. It has a density of 8.9 and is a hard, brittle bluish-white metal, with a melting-point of 1490 °C. It is relatively unreactive and dissolves only slowly in dilute mineral acids. Like the other transition metals, cobalt has several oxidation states, but only states 2 and 3 are of any practical significance. Cobalt(II) forms an extensive series of simple and hydrated salts with all common anions. The hydrated cobaltous (CoII) salts are either red or pink in colour.

Few simple cobaltic salts are known, due to the relative instability of the Co(III) oxidation state. However, numerous complexes of both the cobaltic and the cobaltous forms of the metal exist and are quite stable.

PRODUCTION AND USES

Cobalt does not occur in nature as a simple ore but always in association with other metals, especially with arsenic. The most important ores are smaltite, $CoAs_2$, and cobaltite, $CoAsS$, but the chief technical sources of the metal are actually the residues or speisses resulting from the refining of ores of nickel, copper and lead. The pure metal is separated from speiss by electrolysis.

Total world production is about a quarter of a million tonnes per annum. More than half of this is produced in the two central African countries of Zaïre and Zambia.

The principal use of cobalt is in the metallurgical industry for the production of high-strength alloys. It is also used to make permanent magnets. A small amount of the metal is used in the manufacture of pottery and glass, an application that has been known to man for more than 2000 years. In cobalt-deficient areas of the world, cobalt is added to some fertilisers. A limited quantity also finds application in the pharmaceutical industry.

COBALT IN FOOD AND BEVERAGES

Levels of cobalt in food are very low, unless deliberate addition has been made during processing. The average daily intake has been estimated as being between 5 and 40 μg, according to the reports of several investigators.[54, 55] Concentrations found in different foods range from a high of 1.94 mg/kg in beef fat, to a low of less than 0.05 in white sugar. Vegetables in general contain extremely low and, in many cases, undetectable amounts of the metal. As is the case with some other trace nutrients, refining of raw foodstuffs reduces the levels of the metal in the end-product. Compared to the very low level found, as mentioned above, in white sugar, raw sugar has 0.4 and molasses 0.25 mg/kg of cobalt. Similarly, while whole wheat contains 0.025 mg/kg, white flour has only 0.003 mg/kg, which indicates a loss of almost 90 per cent during milling.[56]

Levels of cobalt in drinking-water are low, usually considerably less than 5 μg/litre. A US domestic supply was reported to have 2.2 μg/litre. Similar levels have been found in untreated surface waters.[57]

Following an incident in which the use of cobalt in beer processing was suspected of having caused heart disease, numerous samples of commercial beer were analysed for cobalt. Levels were well below 0.1 mg/litre, except in the case of those in which cobalt had been used in processing. In these cases up to 1.1 mg/litre of the metal was detected.[58]

ABSORPTION AND STORAGE OF COBALT BY THE HUMAN BODY

There is little precise information available on levels of absorption of ingested cobalt. It is possible that nutritional factors influence absorption and this may account for the fact that estimates of 5–45 per cent absorption have been made by different authors.[59] In rats, 30 per cent absorption of radioactive ^{60}Co, in the form of cobalt chloride, has been reported.[60]

About one-fifth of the total body cobalt appears to be located in the liver, where a concentration of upwards of about 0.05 mg/kg has been found. The total weight of Vitamin B_{12} in an adult liver is about 1.7 mg and this accounts for probably all the cobalt present in the organ. Excretion seems to be mainly in the urine. It is multiphasic, with an initial fast phase followed by a pronounced cobalt retention. As has been noted, cobalt is needed by man only as a component of Vitamin B_{12} and this vitamin must be taken into the body in a preformed state. Only micro-organisms can incorporate cobalt into the vitamin, so that it is the vitamin and not the metal as such that is essential in the diet.

BIOLOGICAL ROLE AND EFFECTS OF COBALT

In 1926 Minot and Murphy discovered that pernicious anaemia, a disease which was until then considered incurable and normally fatal, could be controlled by incorporating large amounts of raw or lightly cooked liver in the diet. Later, an 'anti-pernicious anaemia factor' similar to that of liver was found in the waste liquors from *Streptomyces* fermentations used in the production of antibiotics. This was shown to be a growth factor for several types of micro-organisms, such as those of the rumen of cattle. In 1948 the factor was isolated in the form of red crystals from fermentation broths. Finally in 1956 the chemical structure of this factor, now known as Vitamin B_{12}, was determined.[61]

Vitamin B_{12} was found to be an organic molecule with a high molecular weight of about 1350. Its central structure consists of a ring system rather like that of the porphyrins, made up of four pyrrole units, but with a less extensive system of double bonds than in the porphyrins. This porphyrin-like corrin ring is linked to a nucleotide containing a base, ribose and phosphate. In the centre of the molecule, linked to the four nitrogen atoms of the pyrrole rings, is a single atom of cobalt. The form of Vitamin B_{12} initially isolated had a CN^- (cyanide) group attached to one of the co-ordination positions of the cobalt. Because of this the vitamin was given the name cyanocobalamin. However, it is now generally believed that the cyanide is an addition to the original structure arising during the isolation procedures, and this form of the vitamin occurs rarely, if at all, in nature. Hydroxycobalamin (Vitamin B_{12a}) in which OH^- replaces the CN^- group, does occur in nature. However, the predominant forms are the B_{12} coenzymes, in which an alkyl group replaces the CN^-. Vitamin B_{12} functions physiologically probably in every tissue of the body, but effects of deficiency are most clearly seen where cells are undergoing rapid division, such as in blood-forming tissues of bone marrow. The nervous system is also seriously affected with, sometimes, degeneration of nerve fibres in the spinal cord and peripheral nerves. Normal blood cell formation is also affected when Vitamin B_{12} is deficient.

Abnormal cells known as megaloblasts develop in place of the normal nucleated cells that give rise to red blood corpuscles in bone marrow. The circulating red cells derived from these megaloblasts are bigger than normal, and what is known as macrocytic anaemia results. Megaloblast formation is due to faults in DNA synthesis. Vitamin B_{12} is essential for the formation of the thymine nucleotide, the characteristic base of DNA. Another vitamin, folic acid, is also necessary for the formation of thymidylate.

Vitamin B_{12} itself, and not just cobalt, is essential in the diet if anaemia and the other consequences of deficiency are to be avoided. Less than 5 μg daily are adequate for the normal adult. Unfortunately, anaemia is usually caused not by lack of Vitamin B_{12} but by poor absorption of the vitamin. Absorption depends on an *intrinsic factor*, a mucoprotein, synthesised by the stomach lining. Pernicious anaemia victims sometimes have a hereditary defect which limits production of the factor. Gastrectomy (surgical removal of part of the stomach) can also reduce production of the factor. Pernicious anaemia has also been known to occur in infection with fish tapeworms, which compete with the body for available Vitamin B_{12} and also reduce absorption of it from the gut.

Toxic effects of ingestion of excessive amounts of cobalt have only been reported rarely and then only when cobalt compounds had been deliberately added during food processing or were used in high doses therapeutically for the treatment of certain forms of anaemia. Goitre is a well-known side-effect of cobalt therapy, but usually this effect is reversible.[62] An outbreak of severe heart disease (cardiomyopathy) with a high mortality rate occurred among about 50 beer drinkers in Quebec and Minneapolis in North America and in Belgium in the late 1960s.[63] The outbreak was traced to the introduction of a new brewery process in which cobalt sulphide was added to beer as a foam stabiliser. It was calculated that a heavy beer drinker would take in upwards of 10 mg of cobalt a day from this source (provided he drank 10 litres of beer each day). Since such levels of intake of cobalt have been exceeded in some therapeutic regimes used in the treatment of anaemia without development of heart disease, it has been suggested that a contributory cause of the disease was excessive alcohol intake allied to protein malnutrition. Other health effects have been reported in workers exposed to industrial contamination by cobalt and its compounds. Experimental animals have also been shown to be adversely affected by high intake of cobalt in the diet. However, while levels of cobalt in a normal diet are minute, ingestion of as much as 0.25 mg/day has been shown to have no observable adverse effect on man.[64]

ANALYSIS OF COBALT IN FOOD

A number of colorimetric methods for analysing cobalt in food and other biological materials are available. They are relatively sensitive and have been widely used.[65]

Atomic absorption spectrophotometry permits more sensitive and specific estimations.[66] The air–acetylene flame is suitable and little interference has been reported. However, high concentrations of nickel may cause depression of absorption. This can be overcome by diluting solutions to nickel concentrations below 1500 mg/litre and using a nitrous oxide–acetylene flame. Sensitivity may be increased considerably by extraction of the metal by a sequential procedure before analysis with the atomic absorption spectrophotometer.[67]

NICKEL

CHEMICAL AND PHYSICAL PROPERTIES

Nickel has an atomic weight of 58.71 and an atomic number of 28. Its density is 8.9. It melts at 1453°C and is a tough, silver-white metal. It has high electrical and thermal conductivities and it can be drawn, rolled, forged and polished. Though the metal dissolves readily in diluted mineral acids, it is resistant to attack by air and water and ordinary temperatures and is therefore often used as protective plating on other metals. High oxidation states occur as well as Ni (0) and Ni (I) but only Ni (II) is common. However, is spite of this the chemistry of nickel is not simple because Ni (II) has a tendency to form a wide variety of complexes. These are mainly octahedral, tetrahedral or square in structure, but in addition complicated equilibria, which are generally temperature dependent, may exist between these types.

Divalent nickel forms an extensive series of compounds, many of which are green in colour. A number of binary compounds are formed between nickel and various non-metals such as P and C.

PRODUCTION AND USES

Nickel occurs in nature mainly in combination with arsenic, antimony and sulphur. Among the more important commercial ores are garnierite, a magnesium–nickel silicate, and certain varieties of the iron mineral pyrrhotite, which can contain upwards of 5 per cent nickel.

Elemental nickel is also found alloyed with iron in some meteorites. The central region of the earth is believed to contain considerable quantities of the metal.

Processing of nickel ores and refining of the metal is a complicated process, the details of which depend on the nature of the particular ore being worked. In general, the ore is converted to Ni_2S_3, which is roasted in air to give NiO. This is then reduced with carbon to give the metal. High-quality nickel can be made from lower grade metal by the carbonyl process. In this the crude metal is reacted with carbon monoxide to give volatile $Ni(Co)_4$ and from this metal of 99.99 per cent purity can be produced on thermal decomposition at 200 °C.

Nickel is used in the manufacture of alloys with iron, copper, aluminium, chromium, zinc and molybdenum. Nickel-containing steels are strongly corrosion resistant. It is also used in the production of heat-resisting steels and cast iron. Nickel-plated steels are used in the manufacture of some food-processing vessels and other equipment. An important use of nickel in the food industry is as a catalyst in the hydrogenation of oils.

NICKEL IN FOOD AND BEVERAGES

Nickel is present in small amounts in most soils. Plants may contain between 0.5 and 3.5 mg/kg of the metal.[68] It can be detected in most animal tissues in small amounts. Daily intake in the human diet is about 0.3–0.6 mg/day.[69]

Levels of nickel in the major food classes (in mg/kg) are as follows: cereals, 0–6.45; fruits, 0–0.34; vegetables, 0–2.59; meats, 0–4.5; dairy products, 0–0.03; seafoods, 0.02–1.7.[70]

The average level in most foods is usually less than 0.5 mg/kg, though individual foodstuffs have more. Among these are tea, with approximately 8 mg/kg (dry weight), cocoa (0.98 mg/kg) and nuts (5.1 mg/kg). Seafoods such as herrings and oysters may also be rich in the metal. Higher than average levels of nickel can occur in some foods as a result of contamination during processing. As has been mentioned above, metallic nickel is used as a catalyst in the hydrogenation of oils in the manufacture of margarine. A check on Polish margarine showed that while most samples contained between 0.06 and 0.185 mg/kg, in one type an average of 0.47 mg/kg was found. A sample of Dutch margarine had more than 1.0 mg/kg. These higher than average levels were attributed to insufficient removal of the catalyst from the hydrogenated oil.[71] Similarly, high levels of nickel have been found in gelatine resulting from contamination during processing.

Tobacco has been shown to contain appreciable amounts of nickel. Concern has been expressed that as a result of cigarette smoking nickel carbonyl may be produced in sufficient quantity to cause lung cancer.[72]

The average daily intake of nickel from domestic water supplies in urban reticulated systems has been estimated to be 10 μg. This is based on an average concentration of 4.8 μg/litre.[73] Intake will be higher when larger volumes of fluid are consumed as well as in cases where contamination of water supplies occurs. The use of sewage sludge on agricultural land may be a major source of contamination of water supplies through run-off.

Some samples of sewage sludge have been found to contain between 20 and 5300 mg/kg dry weight of nickel.[74]

METABOLISM AND BIOLOGICAL EFFECTS OF NICKEL

Nickel is poorly absorbed from food and drink. The metal is excreted mainly in faeces, with a smaller amount in urine.[75] Only about 3–6 per cent of dietary intake is retained in body tissues.[76] Distribution in the body seems to be fairly uniform, with no evidence for accumulation to any significant extent in particular organs.[77] It is of some interest that there is a sex-linked difference between the levels of nickel in hair of different subjects. Female hair has been found to contain 3.96 ± 1.055 mg/kg and male hair 0.97 ± 0.147 mg/kg.[78]

Nickel has been shown to be a dietary essential for animals, probably including man. Distinctive deficiency states occur. Chickens fed a nickel-deficient diet grow poorly and have dermatitis and deformation of the legs.[80] Several enzyme systems are activated, though not exclusively, by nickel. These include carboxylase, trypsin and acetyl coenzyme A synthetase. Some of the nickel in human serum is found in a specific nickel-containing protein known as nickeloplasmin.[81]

Nickel is, apparently, non-toxic to man. Though acid foods can pick up the metal from nickel-plated and nickel alloy cooking utensils,[82] it is poorly absorbed and has not been shown to produce toxicity. However, cancer of the respiratory tract as well as dermatitis occur in workers in nickel refineries.[83]

There is a need for more research into the possible chronic effects of small quantities of nickel in the environment. Suggestions that alkyl lead compounds used today as fuel additives should be replaced by organic nickel compounds require careful investigation before the practice is encouraged. In spite of present evidence to the contrary, the possibility that the resulting wide-scale distribution of nickel in the atmosphere and thus on to foodstuffs might have deleterious effects on human health must be taken into account.

METHODS OF ANALYSIS OF NICKEL

Colorimetric methods of analysis, using the coloured complexes formed by nickel with organic chelating agents, have been used widely. Dimethylglyoxime complexes with nickel to produce a red colour which absorbs at 445 nm.

Atomic absorption spectrophotometry is probably the method of choice in today's analytical laboratory. An air–acetylene flame is used and preferably a wavelength of 352.4 nm. This is less sensitive than the 232.0 nm line, but is less susceptible to non-atomic absorption. In hydrochloric and perchloric acid solution, absorbance depression is observed. This can be overcome with, however, reduced sensitivity, by using the nitrous oxide–acetylene flame.

MOLYBDENUM

Molybdenum is the only metal of the second transition series which is of major biological significance. It has long been recognised as an essential nutrient for plants and is almost certainly essential also for animals. However, conclusive evidence for its essentiality for man has not yet been presented; though there is little doubt that the metal plays an important and apparently unique role in the human body. Molybdenum is one of the few heavy metals which appear to be essential for life. Like several of the metals of the first transition series, while it is found in several important enzymes of the body, it can also, if present in more than low concentrations, cause toxicity.

CHEMICAL AND PHYSICAL PROPERTIES

Molybdenum has an atomic weight of 95.94 and is element number 42 in the Periodic Table. Its density is 10.2. The metal is dull silver-white in colour and is ductile and malleable with a high melting-point of 2610 °C. It is strongly resistant to acid attack and is not readily oxidised at ordinary temperatures, but undergoes oxidation on heating in air or oxygen.

The chemistry of molybdenum is complex and is only imperfectly understood. Like the other transition elements, molybdenum is capable of existing in a number of oxidation states. These range from -2 to $+6$. However, in practice only the higher oxidation states are of significance. The lower oxidation states occur mainly in organometallic complexes.

The most stable oxidation state is VI. When the metal is heated in air it is converted into MoO_3. This trioxide is a stable, white solid. It dissolves in alkaline solutions to form molybdates of the form M_2MoO_4. These have a tendency to condense to polymolybdates, $M_6Mo(Mo_6O_{24}).4H_2O$.

Numerous complexes of $Mo(V)$ are known with, for example, cyanides, thiocyanates, oxyhalides and various organic chelates. Similar complexes occur with $Mo(IV)$ and $Mo(III)$. They are all coloured, like the orange-red complex of $Mo(V)$ with CNS^-. The chemistry of the alloys of molybdenum is extensive and complex, as is that of the compounds it forms with non-metallic elements.

PRODUCTION

Molybdenum is one of the rarer elements in the lithosphere and is only the fifty-fourth in the list of abundance. It occurs in soils, generally at a level of 2 mg/kg or less. However, in some areas it is much more abundant and its presence may be indicated by toxic effects on plants and animals. The metal occurs in a number of ores, the most important of which is molybdenite. MoS_2. This is found in a number of parts of the world, but the only major deposit is in Colorado in the USA. Production is about 60 000 tonnes per year. The ore is first concentrated (usually by a foam flotation process) and is then converted to MoO_3 by roasting. This is purified and finally reduced with hydrogen to the metal.

USES

The chief use of molybdenum is in the production of alloy steels. These are extremely hard and strong and are used in the manufacture of machinery and weapons. Molybdenum compounds also find use in the chemical industry as catalysts. Some paints and other pigments are made with molybdenum. A small amount is used in agriculture as a soil dressing.

MOLYBDENUM IN FOOD AND BEVERAGES

Few studies seem to have been carried out on the levels of molybdenum in the human diet, though a lot of attention has been given to the metal in animal feeds. This difference is, no doubt, due to the apparent relative insignificance of molybdenum in human nutrition in contrast to the major problems which have been met due to excessive intake of the metal by farm animals in some parts of the world. Daily intake for man has been estimated to be between 0.1 and 0.5 mg,[84] except where environmental contamination occurs as a result, for example, of mining and metallurgical operations.

The molybdenum content varies between the different classes of foods. Levels are lowest in most meats, root vegetables and dairy products and highest in legumes, cereals, and leafy vegetables as well as animal offal. Beans and other pulses range from 0.2 to 4.7 mg/kg and cereals from 0.12 to 1.14 mg/kg.[85] Molybdenum is concentrated to some extent in the outer layers of the wheat grain, and so levels of the metal will decrease with increasing extraction rates of flour.[86]

Considerably higher than average levels of molybdenum can occur in herbage and in vegetables in areas where the soil is naturally enriched with the metal or where industrial pollution occurs. Levels can also be increased by the use of molybdenum-containing fertilisers or of sewage sludge. The latter may contain upwards of 30 mg/kg dry weight of the metal.[87] In the case of cattle, a distinctive toxic condition known as 'teart disease' occurs when pastures have molybdenum levels up to 100 mg/kg, rather than the normal range of 3–5 mg/kg dry weight.[88]

METABOLISM AND BIOLOGICAL EFFECTS OF MOLYBDENUM

Absorption of molybdenum from the gastro-intestinal tract is high. In humans about 50 per cent of ingested molybdenum enters the blood-stream.[89] However, retention levels appear to be low and most of the molybdenum is excreted by the kidneys in urine.

Molybdenum is a constituent of three intracellular flavoprotein en-zymes: xanthine oxidase, which is involved in the formation of uric acid, aldehyde oxidase and sulphite oxidase. The need for the metal for the formation of xanthine oxidase has been established in animals and significant growth response to molybdenum has been demonstrated. As yet no similar need or response to molybdenum has been demonstrated in man.

Molybdenum metabolism is known to be closely related to that of sulphur and of copper. This is apparently a complex interaction and is not fully understood.

Sulphur and sulphur compounds can in some cases limit molybdenum absorption and retention and increase urinary excretion in animals, and this can alleviate molybdenum toxicity in teart disease. Copper also has an effect on molybdenum absorption and metabolism and generally reduces symptoms caused by an excessive intake of molybdenum.[88] It is of significance that copper poisoning can occur in sheep even though copper intake is moderate when at the same time levels of molybdenum and sulphur in the diet are very low. Furthermore, if copper is low, even moderate amounts of molybdenum can produce toxic effects. These effects are intensified by addition of sulphate. In contrast, when copper is adequate, larger amounts of molybdenum are required to produce molybdenosis. Sulphate can completely prevent this toxicity. To add to the complexity, manganese also has an effect on the copper-sulphur-molybdenum interrelationship.[86]

In the case of humans, urinary excretion of copper has been shown to be affected by molybdenum intake.[89] The existence of the more complex interrelationship involving sulphur and manganese as well as sulphur has not been established.

Only a few reports have been published of molybdenum toxicity in humans. A survey in an area of Armenia in the USSR, where the soil had a relatively high level of molybdenum, indicated a possible relationship between the consumption of locally grown molybdenum-rich vegetables and a high level of incidence of gout. It has been suggested that this local exposure to molybdenum could give rise to an increase in xanthine oxidase activity and a consequent increase in uric acid production leading to gout.[90]

ANALYSIS OF MOLYBDENUM

As with the other transition metals, colorimetric methods of analysis are available for the analysis of molybdenum. Thiocyanate and dithiol have been successfully used and permit analyses to a limit of 0.01 mg/kg dry weight.

Atomic absorption spectrophotometry is used widely today. However, under some conditions, severe interference from other metals has been encountered. To overcome this, a nitrous oxide–acetylene flame is recommended. Addition of an excess of a refractory element (for example, $1000\,\mu g/ml$ of aluminium) also helps to reduce interference.[91]

10

Other Transition Metals

The only other metals of the three transition series which are of any real significance as possible contaminants of food are titanium and vanadium of the first series, silver of the second and tungsten of the third. Since the probability of such contamination occurring or at least of its having any major effect on the human organism is low, these metals will be considered only briefly.

TITANIUM

CHEMICAL AND PHYSICAL PROPERTIES

Titanium has an atomic weight of 47.9, and is number 22 in the Periodic Table. It is a hard, though light (density 4.5) metal, with a melting-point of 1668 °C. It is strongly resistant to corrosion. Titanium is the eighth most common element, making up about 0.6 per cent of the earth's crust. It is produced commercially from two major minerals, ilmenite and rutile. World production is about $1\frac{1}{2}$ million tonnes.

Titanium metal is used extensively in the aerospace industry because of its lightness, toughness and resistance to oxidation. Titanium oxide is used frequently as a white pigment in paints, enamels, paper coatings and plastics.

In the food industry titanium oxide is used as an additive for whitening flour, confectionery and dairy products. It is also used in dry beverage mixes.

Pharmaceutical use is made of titanium as an ultraviolet screen in suntan lotions.

TITANIUM IN FOOD AND BEVERAGES

Because of its wide-scale distribution, titanium is an almost universal contaminant of vegetables, cereals and other plant foodstuffs. However, this is mainly external contamination, for the metal is poorly absorbed and retained by plants and animals.[1] Levels of about 1.8–2.4 mg/kg have been recorded in cereals, vegetables, dairy products and other foods. Domestic water supplies contain about 2 μg/litre of titanium on average. Daily intake has been estimated to be between 3 and 600 μg in the USA.[2] It is possible that intake might be greater in the case of people living in the vicinity of coal- or oil-fired power plants.

ABSORPTION AND METABOLISM OF TITANIUM

Few studies appear to have been carried out on the metabolism of titanium by the human body. What evidence exists points to a very low level of gastro-intestinal absorption. Excretion of absorbed titanium seems to be rapid, but experimental data is lacking on excretory rates and routes. No evidence that absorbed titanium performs any vital function or that it is a dietary essential has been produced. Similarly no evidence points to toxic effects of titanium or of its compounds in food.

METHODS OF ANALYSIS OF TITANIUM

Several instrumental methods of analysis for titanium can be employed. These include X-ray fluorescence, neutron activation analysis, polarography and atomic absorption spectrophotometry. This latter method uses a nitrous oxide–acetylene flame. Interference is experienced from many other metals. Most metals enhance the titanium absorption signal, but sodium and several anions reduce it.

VANADIUM

CHEMICAL AND PHYSICAL PROPERTIES

Vanadium has an atomic weight of 50.9 and is element number 23 in the Periodic Table. It is a relatively light metal with a density of 6.1. The pure metal is corrosion resistant, hard and steel-grey in colour. However, the metal is mainly used in the form of an iron alloy called ferrovanadium, and not as the pure metal. The alloy is used in the production of steel and cast iron to which it confers shock resistance and ductility.

Vanadium, like other transition metals, can assume a number of oxidation states from $+2$ to $+5$. The vanadate ion VO_4^{3-} is the predominant form of V(V) in basic solution. The VO^{2+} ion is an especially stable unit in compounds of V(IV). The chemistry of vanadium suggests that it can have a redox role in biological systems corresponding to that of iron and copper.

PRODUCTION AND USES

Vanadium is widely distributed, with a concentration of about 0.02 per cent in the lithosphere. However, simple ores are rare and it usually occurs in combination with other metals. The mineral carnotite, for instance, contains uranium, for which the ore is worked, as well as vanadium, which may also be recovered.

Extraction of pure vanadium from ores is difficult, because of the reactivity of the metal at high temperatures. For this reason, as has been mentioned, commercial production is mainly of ferrovanadium alloy. The main user of vanadium and its alloys is the metallurgical industry.

VANADIUM IN FOOD AND BEVERAGES

Little information is available about the levels of vanadium in the human diet. Results reported show considerable variation. This may reflect unreliability of the analytical methods used. Levels in different food groups have been reported to be as follows,[3] (in mg/kg fresh weight): seafoods, 0–51; cereals, 0–6.03; fruits, 0–0.18; vegetables, 0–6.0; nuts, 0–1.96; with none in meats. The average daily intake has been estimated to be 2.0 mg, but this will depend on the pattern of foods eaten. It has been suggested that foods rich in fats and oils will contribute higher levels of vanadium to the diet. Vegetable oils such as soya, corn, groundnut and olive have been reported to contain over 40 mg/kg of vanadium.[4]

It is possible, also, that intake of vanadium might increase as the result of the use of sewage sludge in agriculture. Sludge has been shown to contain between 20 and 400 mg/kg of the metal.[5]

ABSORPTION AND METABOLISM OF VANADIUM

Vanadium is poorly absorbed from the gastro-intestinal tract. Retained vanadium accumulates to some extent in bones and teeth. It is found in human tooth enamel and dentine and may exchange with phosphorus in the apatite tooth substance.[6]

Vanadium has been shown to be an essential nutrient for rats and some other animals. In the marine organisms, the tunicates or sea-squirts, vanadium apparently plays a biochemical role, possibly as an oxygen carrier, in vanadocytes—green blood cells containing 4 per cent of V(III) and about 2N H_2SO_4. No similar role for vanadium has been established in higher organisms. However, its universal presence in tissues of animals, including man, suggests that it is also needed by them. It may function in lipid metabolism. High doses of vanadium may inhibit cholesterol synthesis and lower the phospholipid and cholesterol content of the blood. The metal may also inhibit development of caries by stimulating mineralisation of teeth.[8]

Vanadium does not appear to be toxic to man, though it apparently has toxic effects on rats at fairly low dietary levels. Industrial exposure of man to vanadium dusts has been shown to cause, in addition to eye and lung irritation, inhibition of activity of the enzyme cholinesterase, resulting in a deficiency of choline. Such a deficiency may have adverse effects, including liver and kidney damage. However, there are no reports of any such problems arising from dietary intake of vanadium.

ANALYSIS OF VANADIUM

Vanadium in acid-digested food samples may be determined by atomic absorption spectrophotometry. A nitrous oxide–acetylene flame is employed. There is some interference by fluorides, ammonia and some other ions. This interference can be removed by the addition of aluminium (2000 μg/ml) to the test solution.

SILVER

CHEMICAL AND PHYSICAL PROPERTIES

Silver is element number 47 in the Periodic Table, with an atomic weight of 107.87. It is a white, lustrous, soft and malleable metal, capable of a high degree of polish. Its melting-point is 960 °C. It has the highest known electrical and thermal conductivities—two properties which account for its use in the electrical industry as well as in the gourmet kitchen.

Silver is relatively unreactive, though it does blacken in the presence of sulphur and hydrogen sulphide. Though oxidation states I and II are known, the normal and predominant oxidation state of silver is Ag(I), the argentous. Silver compounds of importance are the Ag(I) nitrate, acetate, and halides.

PRODUCTION AND USES

Silver has long been produced as a by-product of the lead-refining industry. The cupellation of silver-containing lead ores to give pure silver has been practical since Roman times and was the main source of the metal in the mines of Britain.

Silver is also produced on a large scale from argentite. The world's main producers of silver are Canada, Peru, the USSR, the USA and Mexico.

Apart from its use in jewellery and coins, silver is used in tableware. It forms a number of valuable solders and alloys with copper, cadmium and lead. It is widely used chemically in the photographic industry. Silver salts have germicidal properties and are used in water treatment. There are a number of pharmaceutical applications of silver.

SILVER IN FOOD AND BEVERAGES

Little information is available on levels of silver ingested in the diet. Intake is probably not more than about 20 μg/day.[9] Levels in cooked food may be increased if silver or silver-plated cooking and eating utensils are used regularly, but even this additional contribution is probably very low.[10] Drinking-water normally seems to contain very little silver except when it has been treated with silver for disinfection purposes.[11] To prevent high intakes from this source the US Drinking Water Standard (1962) for silver is set at a maximum of 50 μg/litre. There is some indication that vegetables and other vegetation may accumulate higher than average levels of silver in the vicinity of coal-burning power-stations.[12] It has been suggested that cabbage and other vegetables of the *Brassica* genus are particularly likely to accumulate silver under such conditions, as well as from silver-containing water.

ABSORPTION AND METABOLISM OF SILVER

Little is known about the levels of absorption of silver in the gastro-intestinal tract. Evidence from animal studies indicates that absorption may be as low as 10 per cent. Accumulation of absorbed silver seems to be in liver as well as in skin.

Silver has not been shown to be essential for the human organism. In animals, ingestion of silver nitrate at levels of several hundred milligrams a day has resulted in cardiovascular and liver toxicity.

Prolonged ingestion of silver salts in pharmaceutical preparations by humans can bring about a permanent blue-grey discoloration of the skin, eyes and mucous membranes, known as argyria. Clinical symptoms may also occur in the lungs and other organs.[13] The consumption of 'antismok-ing lozenges' has been reported to cause such an effect.[14] There are no reports of similar effects caused by silver ingested in food. The presence of selenium as well as of copper and vitamin E in the diet has been shown to decrease the toxicity of silver to turkeys.[15]

ANALYSIS OF SILVER

A dithiozone colorimetric method can be used for the estimation of silver. A polarographic method allows considerably lower levels of detection. Both atomic absorption and neutron activation analysis are also used. No chemical interference has been observed in atomic absorption spectropho-tometry when an air–acetylene flame is used.

11

Zinc—The Unassuming Nutrient

There are good reasons for describing zinc as 'the unassuming nutrient'.[1] Indeed, in spite of the important role the metal has long played in man's biological and economic life, it has never attracted very much attention from scientists. It comes as a surprise to many to learn that the human body needs, and contains, almost as much zinc as it does iron, and more than 10 times more than it does of copper, two metals that are well known as essential inorganic nutrients.

Some 80 enzymes have been shown to contain zinc. In the absence of an adequate dietary supply of the metal serious health problems arise. To some extent the overlooking of the importance of zinc by biochemists and medical scientists is due to its physico-chemical properties. Unlike the transition metals, zinc does not form many distinctive and brightly coloured complexes with chelating agents. Neither does zinc, though it is an industrially important metal, play a major structural role in construction or manufacture. It has never had the distinction of having an age of human history named after it, as had iron and the copper alloy bronze. However, today there is a growing interest in the role of zinc in human health and at the same time a steady increase in world production of the metal. The very important place of this metal in human health is now generally recognised. The supreme accolade was paid, at least by British nutritionists, when the metal joined the ranks of iron and copper in the tables of composition of foods in the latest edition of McCance and Widdowson,[2] the nutritionist's 'bible'.

CHEMICAL AND PHYSICAL PROPERTIES

Zinc has an atomic weight of 65.37 and is number 30 in the Periodic Table. Its density is 7.14, with a melting-point of 420 °C. Like mercury and, to a lesser extent, cadmium, its two neighbours in the Group IIb elements, zinc is remarkably volatile for a heavy metal. It is a bluish-white, lustrous metal which is ductile and malleable at 100 °C. It tarnishes in air to a blue-grey colour due to the formation of an adhering coat of a basic carbonate, $Zn_2(OH)_2CO_3$. This layer protects the underlying metal from further corrosion and is the basis for the use of zinc in galvanising other metals to protect them from corrosion.

Zinc is a reactive metal. It combines readily with non-oxidising acids, releasing hydrogen and forming zinc salts. It also dissolves in strong bases to form zincate ions, $(ZnO_2)^{2-}$. It reacts with oxygen, especially on heating, producing zinc oxide. It also reacts directly with the halogens and with sulphur and other non-metals. Though not strictly speaking a transition metal, since it has a completely filled *3d* electron shell, it does share with copper and the other transition metals a tendency to form strong complexes with organic ligands. It is no doubt this property which has contributed to the important biological role assumed by zinc during the course of evolution.

Zinc forms a number of alloys, the best known and most important of which is brass. This alloy has been produced and used by man for more than 2000 years and is still of considerable importance. Bronze also contains zinc, usually in small amounts.

PRODUCTION AND USES

Zinc ores are widely distributed and mines are operated in several European countries as well as on a larger scale in the USA, USSR, Southern Africa and Australia. The principal ores are sulphides, such as zinc blende. The carbonate *calamine*, oxide *zincite*, and silicate *willemite*, are also worked commercially. The metal frequently occurs in association with other metals such as lead, cadmium and copper.

Ores are crushed and may be concentrated by, for example, a wet flotation process. The concentrate is then roasted to produce the oxide, which is subsequently reduced, usually with coke, and the metal distilled off. During smelting large emissions of volatile zinc may occur. As a consequence a considerable environmental hazard may result, especially if cadmium is also present in the ores. Emission problems have been overcome in some operations in which the traditional imperial smelters have been replaced by rotary kilns.

World production of zinc was about 5 million tonnes in 1975.

Consumption will be about 5 500 000 tonnes in 1980.[3] Recovery of zinc from scrap accounts for about one-sixth of the metal used annually in the USA.

The principal use of zinc for many centuries was in the production of brass. This alloy was produced in large quantities in China and India and was an important item in trade when European ships began to penetrate the Pacific Ocean. Today, brass is still of importance and continues to be used, as it was traditionally, for the manufacture of food-processing and catering equipment. However, apart possibly from the Middle East and other parts of Asia, culinary use of brass has declined with the advent of modern cooking equipment.

A major use of zinc is to protect iron and other metals from corrosion by air and water, in the form of galvanised steel and iron. The zinc is applied by various methods such as hot dipping or by dusting (sherardising) as a thin layer on the surface of the metal to be protected. The external zinc undergoes corrosion to form a protective skin of basic carbonate. The zinc also protects less active metals such as iron because it corrodes first as a more active, 'sacrificial' metal. About one-half of the zinc produced is used in this way, much of it in the motor industry. In 1973, for instance, out of the total of $1\frac{1}{2}$ million short tons of zinc used in the USA, half a million were consumed in the motor industry.[4]

Zinc oxide is used in large quantities in the manufacture of rubber and as a white pigment. A considerable amount is also used in the manufacture of dry cells.

Zinc is important in the pharmaceutical industry, where it is used in nutrient supplements, ointments, shampoos and other preparations. The carbonate has been used as a pesticide. Organic zinc compounds can replace alkyl lead in antiknock petroleum additives. Zinc compounds are also used in a number of industrial operations, such as the manufacture of fibreboard.

ZINC IN FOOD AND BEVERAGES

Zinc is widely distributed in food and beverages, though concentrations may be low, especially in processed commodities of vegetable origin. In general, the richest dietary sources of zinc are seafoods. Oysters, for example, on average range from 60 to 1000 mg/kg. Raw mackerel has 5 mg/kg and a traditional British fish dish—plaice fried in breadcrumbs—has 10 mg/kg. Meat is another good source, with raw lean beef containing

33, raw lamb 29 and raw pork 18 mg/kg. Nuts are also relatively rich in zinc. Brazil nuts contain 42 and salted roast peanuts 30 mg/kg. Vegetables are generally poorer sources of zinc. Uncooked green (Savoy) cabbage has 3 mg/kg. This falls to 2 mg/kg in cooking. Boiled potatoes also contain 2 mg/kg, though the level rises somewhat, to 6 mg/kg, in cooked chips. Fruit are also poor sources of zinc, with 1 mg/kg in eating apples and 2 mg/kg in a fresh orange. Cereals which are initially relatively good sources of zinc, lose some of the metal during milling. Wholewheat flour (100 per cent) has 30 mg/kg, but plain white flour only 6 mg/kg. Polished rice has 13.7 mg/kg of zinc compared to 16.4 mg/kg in the unpolished grain.[5] Dairy products are also fairly low in zinc. The range reported in fresh whole cows' milk is 2–6 mg/litre. Butter contains 1.5 mg/kg.

These figures for zinc in food, which mainly relate to UK foodstuffs,[2] are in agreement with levels reported from the USA[6] and other industrialised countries. The daily intake of zinc by 'reference man' in the UK is calculated[7] to be 13.0 mg and in the USA the diet contains between 10 and 15 mg.[8] The intake per head of the metal in the Netherlands is 16.8 mg, with variations between different groups of the population of 15.4–24.6 mg/day.[9] Since the recommended daily allowance of zinc in the USA is 15 mg/day for an adult, with an additional 5 mg for women during pregnancy and 10 mg during lactation, the figures for actual intake appear to indicate that, apart from the Netherlands, the normal diet may not always meet human requirements. There is, indeed, some indication that among particularly susceptible groups such as children, where demand for the metal is relatively high because of the requirements for physical growth, chronic zinc deficiency may be fairly common.

Natural water generally contains about 10 μg/litre,[10] but reticulated water may have much more, depending on the nature of the plumbing system. An average level of 194 μg/litre has been reported for domestic water in one US city.[11] Contamination of domestic water supplies by sewage sludge, which can contain between 700 and 49 000 mg/kg of zinc, can also result in very high levels of zinc intake in the diet.[12]

ABSORPTION AND METABOLISM OF ZINC

The gastro-intestinal absorption of zinc is affected by a number of factors and is highly variable. Uptake has been reported to range from less than 10 per cent to over 90 per cent, with an average of between 20 and 30 per cent or possibly higher.[13] To some extent uptake seems to be regulated by the zinc status of the body. In addition, various components of the diet have been shown to interfere with zinc absorption. These include phytic acid, fibre and calcium. The restriction on zinc intake resulting, for example, from a diet rich in wholewheat bread, which contains all three of these components, may be so severe as to result in symptoms of zinc deficiency.[14]

There is also some evidence of competition for uptake between zinc and cadmium.[5] The bioavailability of zinc appears to be greater in foods of animal than of vegetable origin. It has been suggested that this availability may be related to the presence in some foods of zinc-binding ligands which facilitate uptake. The greater availability of zinc from human compared to cows' milk has been attributed to the presence of such a ligand.[15]

Absorbed zinc is probably bound during its passage through the mucosal cell to a low molecular weight ligand secreted by the pancreas.[16] In the blood the metal probably binds to albumin or transferrin. Intestinal absorption seems to be related to the level of metallothionein, which sequesters excess zinc in the mucosal cells.[17] Absorbed zinc accumulates rapidly in the liver, pancreas, spleen and kidneys. Highest concentrations are found in the prostate gland (which may contain as much as 100 mg/kg) and also in the eye. Other organs and muscles contain about 25–50 mg/kg. Zinc accumulates in hair and nails and these tissues have been used as an indicator of dietary uptake.[18] Levels in blood plasma are normally about 1 mg/litre.[19]

Zinc is largely excreted via the faeces. These contain both unabsorbed metal as well as that secreted into the intestine, primarily by the pancreas. Only a small proportion of total zinc excreted appears in urine, with a tiny fraction in sweat.

BIOLOGICAL EFFECTS OF ZINC

Zinc is now well established as a dietary essential for man and recommended daily allowances have been established in many countries. It plays a number of important biological roles, especially in several enzyme systems. One of the most important of these is carbonic anhydrase, which is involved in the transport of carbon dioxide in the blood and its release into the lungs during respiration. Zinc enzymes occur in each of the six categories designated by the IUB Commission on Enzyme Nomenclature. These include nucleic acid polymerases, lactic acid, alcohol and retinol (vitamin A) dehydrogenases, as well as some phosphatases, proteases and others. Zinc deficiency manifests itself in numerous symptoms which undoubtedly include signs of defective operation of these enzymes.

The metal also plays an important role in biological structure. It operates for example, in the β cells of the pancreas, where it appears to stabilise the insulin molecule. It functions, probably in a similar manner, in the choroid region of the eye in binding the retina in position.

An inadequate supply of zinc in the diet of young animals rapidly gives rise to several clinical symptoms. These include retardation of growth, poor or erratic consumption of food and loss of appetite, alopecia and the development of scaly keratinous lesions on the skin. In addition there is evidence that zinc deficiency reduces resistance to infection and slows wound healing after injury. Conditions corresponding to these have also been observed in humans. An extreme example is the inherited disease acrodermatitis enteropatica, in which there is an inability to absorb zinc from the diet. Some similar symptoms also occur in states of 'conditioned deficiency', where diets rich in factors which decrease bioavailability of zinc result in dwarfism and failure to develop sexually. It is possible that less obvious symptoms of chronic suboptimal zinc deficiency may be seen more frequently in the future as the consumption of highly refined, low zinc foods increases.[1]

TOXIC EFFECTS OF ZINC IN FOOD AND BEVERAGES

Zinc salts in toxic levels act as gastro-intestinal irritants and have been reported to result in an acute but transitory illness, with symptoms of nausea, vomiting, abdominal distress, cramp and diarrhoea.[20] However, it should be remembered that cadmium is frequently present along with zinc and it is possible that symptoms that have been attributed in the past to zinc poisoning may actually have been indications of cadmium toxicity. In fact, reports of toxic effects following the ingestion of zinc are uncommon and, compared to most other trace metals, zinc is fairly non-toxic.

The few cases that have been reported in the literature relate mainly to the use of galvanised iron vessels. One case involved approximately 300 people out of a group of some 400 who attended a party in which food which had been held overnight in galvanised vessels was eaten. Symptoms were as mentioned above and appeared 3–10 hours after eating and lasted in most cases for about 18–24 hours. The emetic dose of zinc in this case was estimated to be between 225 and 450 mg, equivalent to about 1–2 g of zinc sulphate. Another case involved the poisoning of a group of people who drank an alcoholic punch held in a galvanised container for just over two days before consumption. The beverage was found to contain 2200 mg/litre of zinc.

Levels of over 100 mg/litre of zinc have been reported to be commonly present in some home-brewed beers and other beverages in the preparation of which non-food-grade metal containers had been used.[21] However, it is not known if chronic zinc toxicity resulted from the continuing ingestion of such beverages. Prolonged oral treatment with zinc salts of patients suffering from leg ulcers does not appear to have resulted in chronic zinc poisoning, pointing to the low toxicity of this metal. Nevertheless, it is undesirable that galvanised containers should be used for the preparation or storage of food and beverages, especially when these are of low pH, because of the possibility of solubilisation of the metal.

METHODS OF ANALYSIS OF ZINC

The dithiozone colorimetric method for the detection and determination of zinc is still frequently used. The coloured complex formed with the chelating agent is extracted into an organic solvent and compared with standard solutions of zinc prepared in the same manner. The detection limit is about 0.7 mg/litre.

The most widely used method today is atomic absorption spectrophotometry. Few problems arise with this method. It is sensitive and there is little interference from other elements.

12

Beryllium, Strontium, Barium and the Other Metals—Summing Up

The metals which have not so far been treated are not of major importance as contaminants of food. However, under special circumstances they can get into food and the consequences for the consumer, as well as for the manufacturer, may in some cases be considerable. With the increasing use of several of these once rare metals, often in connection with the space industry and its technological 'spin-off', the likelihood of contamination of food is constantly growing. It will be useful, therefore, to look briefly at some of these potential contaminants from the point of view of present uses and in anticipation of future developments.

BERYLLIUM

Beryllium belongs, as do strontium and barium, to Group IIA of the Periodic Table, along with magnesium, calcium and radium. These metals are known collectively as the alkaline earth metals. Beryllium is element number four in the Table, with an atomic weight of 9.0 and a density of only 1.85. It is a light, silvery white, extremely strong, flexible metal with a number of important industrial uses. It is used to make windows in X-ray tubes, because it is the most penetrable of the light metals to X-rays. Special alloys of beryllium with, for example, copper are used to make springs subject to frequent vibration and also in equipment requiring high electrical and thermal conductivity. Beryllium alloys also have the considerable advantage that they are very strong, do not spark and are non-magnetic. Beryllium oxide was formerly used in fluorescent tubes and screens, but this is no longer done since the extremely toxic nature of the oxide and all other beryllium compounds was recognised. The industrial use of beryllium is constantly growing and though the amount used in individual items may be minute, its alloys are present in many products in daily use. In addition, beryllium and its compounds play important roles in the nuclear, space and military equipment industries. Chemically, beryllium resembles aluminium. For example, its salts are readily hydrolysed in water, and the hydroxide, $Be(OH)_2$, is amphoteric. When the hydroxide is dissolved in alkali, compounds of the type Na_2BeO_2 are formed.

PRODUCTION AND USES

Beryllium is an extremely costly metal to produce. The main source is the aluminosilicate gemstone beryl. One ton of beryl produces 35 kg of beryllium. It also occurs in a number of other gems such as the emerald. Beryl is first converted to the hydroxide by roasting and leaching, and then to a halide, and either the pure metal produced by reduction of this with magnesium or the halide is melted with copper to form an alloy. Because of its cost and also the fact that it is only used in tiny amounts in the production of alloys, its main consumer, world production of beryllium is measured in kilograms, not in tonnes.

BERYLLIUM IN FOOD AND BEVERAGES

Little information is available on the normal levels of beryllium in the diet. It has been estimated that daily intake in an industrial society is about 100 μg. Levels in food found in one study were: potatoes, 0.17; tomatoes, 0.24; rice, 0.08; bread, 0.12 mg/kg (dry weight). Little naturally occurring beryllium has been reported in vegetation or in water, but environmental contamination has occurred in the neighbourhood of coal-burning power-stations and of some metallurgical industries.[2]

Some plants have been reported to accumulate beryllium and these could be a potential source of high beryllium intake for stock as well as man. Cigarette smoke may also contain beryllium. Since beryllium alloys are used increasingly in electrical switches and relays, components of instruments, moving parts of equipment and motors of many kinds, concern has been expressed that situations might arise in the food-processing industry where beryllium-containing equipment was used in duties for which it was not originally intended. Because of the highly toxic nature of the metal, the consequences could be very serious.[3]

METABOLISM AND BIOLOGICAL EFFECTS

Animal experiments indicate that gastro-intestinal absorption of dietary beryllium is low, with about 20 per cent being taken in under the acid conditions of the stomach.[4] Absorbed beryllium appears to be carried in all tissues and organs as a colloidal phosphate bound to protein. Some of the beryllium seems to act in a manner similar to magnesium and becomes incorporated into bone. The dangerously toxic properties of beryllium were first recognised on a large scale during the 1940s, when the term beryllosis was given to an illness which followed soon after the inhalation of dust and fumes of salts of the then relatively obscure metal beryllium.[5] Hundreds of workers, including some in the newly established nuclear industry, who used the metal, its alloys and salts, developed the illness and several died. It was found that particles of the metal or its compounds, wherever they were deposited, but especially in the lungs and skin, caused extensive local damage to the tissues.

One particularly unfortunate source of beryllium poisoning turned out to be fluorescent tubes which were coated with beryllium phosphors. This use has been abandoned now for some 30 years, but old tubes may still be a hazard if broken pieces become embedded in the skin. Beryllium toxicity is not yet fully understood. The metal is a powerful inhibitor of several enzymes, in particular of alkaline phosphatase. It also affects protein and nucleic acid metabolism, possibly leading to the development of cancer. Other effects at the cellular level have been reported. Inhalation of beryllium dusts and physical contact account for the cases of beryllosis reported in the literature. There is as yet no evidence that ingestion of the metal or its compounds has resulted in human illness, even in what are known as 'neighbourhood cases'. These are outbreaks of beryllosis, not among the workers, but among people living in the vicinity of a beryllium plant. Nevertheless, early studies with rats showed that ingested beryllium salts could have serious effects on bone and marrow structure.[6] Whether such effects could result from the ingestion of food which had been contaminated by contact with beryllium or its salts has not yet been established. However, because of this possibility, the use of beryllium alloys in equipment used for food processing is not recommended.

METHODS OF ANALYSIS

Colorimetric methods using reagents such as naphthochrome green G and alkannin are useful but require prolonged preparatory procedures.

Atomic absorption spectrophotometry methods have been reported. A nitrous oxide–acetylene flame is usually used. Severe depression of absorption is caused by sodium and silicon in excess of 1 mg/cm^3. Depression is also caused by aluminium, but this can be overcome by incorporating fluoride into the test solution.[7]

The most sensitive method for beryllium analysis is gas chromatography coupled with an electron capture detector[8] or a mass spectrometer.[9]

STRONTIUM

We have already looked at the radioactive isotope of this element, ^{90}Sr, a potentially hazardous by-product of nuclear fission. Here, as a contrast, we will consider briefly non-radioactive strontium which, if it is a health hazard at all, is only a minor one.

CHEMICAL AND PHYSICAL PROPERTIES

Strontium is element number 38 in the Periodic Table with an atomic weight of 87.62. It is a silvery white, malleable and ductile metal with a melting-point of 768 °C. Its chemical properties are similar to those of calcium.

PREPARATION AND USES

Strontium is a relatively abundant metal in the lithosphere. It occurs, for example, in sea-water at a level of about 8 mg/litre.[10] It is extracted from its chief ores, celestine, $SrSO_4$, and strontianite, $SrCO_3$, by electrolytic procedures. World production is about 12 000 tonnes per annum. The metal is used mainly for its chemical and not its metallic properties. It serves as a 'getter' to remove traces of gas from vacuum tubes. Strontium salts provide some of the brilliant colours of fireworks and flares. They are also used in ceramics and plastics. Small amounts of the metal are used in purifying zinc, in permanent magnets and in iron castings. A small amount is also used in the pharmaceutical industry.

STRONTIUM IN FOOD AND BEVERAGES

Little work has been reported on non-radioactive strontium in food. Levels in plants growing on normal soil range from 1–169 mg/kg dry weight.[11]

However, certain plants have the ability to accumulate the element, sometimes up to several thousand milligrams per kilogram. Levels in animal tissue range from about 0.06 to 0.50 mg/kg. The daily dietary intake of an adult has been reported to be between 0.4 and 2.0 mg.

METABOLISM AND BIOLOGICAL EFFECTS

Strontium is poorly absorbed in the gastro-intestinal tract. Most of what is eaten is excreted directly in the faeces. What is absorbed probably competes with calcium and some is deposited in bone. The ingestion of large amounts of strontium lactate over a period of more than a year by mice was found to inhibit calcification in growing bones and resulted in stunted growth.[12]

Injections of strontium chloride into rats resulted in respiratory failure and death.[13] In the case of man, however, no evidence has been produced indicating toxic effects of ingested non-radioactive strontium.

ESTIMATION OF STRONTIUM

Strontium is readily estimated by both emission and atomic absorption spectrophotometry. A nitrous oxide–acetylene flame is used in both methods. In an air–acetylene flame, silicon, aluminium, titanium, zirconium, phosphate and sulphate depress the signal at all concentrations. Addition of lanthanum salts or of EDTA will overcome these effects.[14]

Strontium is partly ionised in all flames. Addition of potassium chloride or nitrate to all solutions will suppress this ionisation.

BARIUM

Barium is not likely to be a common contaminant of foodstuffs. However, its use in domestic products as well as in medicine, and at least one reported case of widespread poisoning due to contamination of salt with the metal, suggest that barium should be considered briefly here.

CHEMICAL AND PHYSICAL PROPERTIES

Barium, element number 56 in the Periodic Table, with an atomic weight of 137.3, is the heaviest of the alkaline earth metals. It is a silvery yellow metal with a melting-point of 850°C and density 3.7. Its chemistry is similar to that of calcium in many respects. It is of significance, in the present context, that barium sulphate is one of the least soluble compounds known.

PRODUCTION AND USES

The main source of barium is the naturally occurring sulphate, barytes. This is usually reduced with charcoal to the sulphide, which is then used to produce the various chemical compounds which are the forms in which most of the metal is used industrially. About 4 million tonnes of barium are produced annually, mainly in the USA, the USSR, Canada, Mexico and Germany. Smaller amounts of barytes are mined elsewhere for direct use without further processing in oil and gas drilling. About 80 per cent of all barytes produced today is used in this way as drilling muds. Barium compounds are used in the manufacture of glass, as fillers for paper, textiles and leather, in ceramics, television tubes, bricks, as pigments in paint, in lubricating oil additives and in the manufacture of permanent magnets. Some is also used in sugar refining and also in a number of consumer products. These include insecticides, rodenticides and various cosmetic products such as depilatories. An important use of barium sulphate is as an X-ray contrast material for gastro-intestinal examinations.

BARIUM IN FOOD AND BEVERAGES

Levels of barium in food show a wide variation in individual items, ranging from 10 to several thousand mg/kg.[15] The average daily intake of the general population has been estimated to be up to 1.33 mg.[16] A vegetarian diet, especially if nuts form a significant part of it, could be particularly high in barium. Some marine organisms can concentrate the element from sea water. Similarly among vegetables, soya beans and tomatoes have been found to accumulate the element from soil 2–20 fold.[17] Barium levels in domestic water supplies in the US range from 7 μg to 15 mg/litre.[18]

METABOLISM AND BIOLOGICAL EFFECTS

Barium is poorly absorbed from food and there is generally little retention by tissues. As has been mentioned, barium sulphate is especially insoluble and there is no evidence that pre-X-ray 'barium meals' result in any gastro-intestinal absorption of the compound. Soluble salts of barium have been shown to be absorbed to a level of about 10–30 per cent in hamsters.[19] When absorbed, barium has a tendency, like all the other alkaline earth metals, to accumulate in the skeleton.

There is no evidence that barium performs an essential role in the human body. Some earlier reports suggested an essential role in plants and animals, but these findings have not been substantiated.[20] Accidental and suicidal poisonings with barium-containing household and medicinal products have occurred. An outbreak of food poisoning due to natural contamination of locally mined table salt in China has been reported.[21]

METHODS OF ANALYSIS OF BARIUM

Colorimetric methods for microquantities of barium are unsatisfactory. An atomic absorption procedure is sometimes adopted. However, a number of difficulties are experienced with the method. Barium absorbance in an air–acetylene flame is severely depressed by phosphate, silicon and aluminium. Use of a nitrous oxide–acetylene flame overcomes the difficulty, but in this flame barium is partially ionised. Addition of a solution of potassium nitrate or chloride is necessary to suppress the ionisation. In addition, the strong emission of barium results in a noisy signal which becomes worse with increasing concentrations. The most satisfactory method at present available is neutron activation analysis.

THE OTHER METALS—SUMMING UP

Apart from the 22 metals which have been discussed at different lengths depending on their prominence and the likelihood of their being con-taminants, all of the remaining 60 or so metals could also be found in food. Some of them, such as boron, tellurium and zirconium, because of their relative abundance and wide-scale distribution, are always present in food, often in appreciable amounts. Others will be present normally only as traces. Among these metals are those which are toxic, at least at certain concentrations, and others which are considered to have no biological effects, good or bad. At least one is known to be an essential nutrient for plants and possibly for animals. It is likely that others will also be seen as essential nutrients as our analytical and experimental techniques continue to improve. Indeed, it just as likely that our growing knowledge of toxicity will also tell us that some of the metals we at present consider to be physiologically inert are, in fact, toxic and the cause of chronic poisoning.

We cannot look at more than a very small number of these remaining metals, and our consideration will have to be brief. It will be sufficient to indicate points of interest and leave it to future volumes to expand on the details when, and if, further research indicates that fuller treatment of certain metals is called for.

BORON

Boron (atomic weight 10.8; atomic number 5) is a widely distributed and, from the point of view of agriculture, biologically important metal. It occurs in deposits mainly in the form of its sodium salt, borax. It has industrial uses in steel production, in the manufacture of electric wires, in glass, enamel and ceramic making. Boron and borax are used in water softeners, fertilisers and in pharmaceuticals.

Upwards of 20 mg of boron may be consumed each day in a vegetarian diet, since plant foods are especially rich in the element.[22] It can also enter the gastro-intestinal tract as a result of accidental intake of pharmaceutical and domestic products which contain it. Boron is readily absorbed in the gastro-intestinal tract. Normally excretion is rapid in urine and there is little retention in tissues.

Though boron is an essential nutrient for plant growth, its essentiality for animals has not been established and no deficiency symptoms are known in humans. A high intake of the element from pasture on boron-rich soil has been known to cause gastro-intestinal and pulmonary disorders in lambs. Serious poisoning has occurred in humans through accidental intake of large amounts of boron compounds.[23]

BISMUTH

Though not employed industrially in large quantities, bismuth (atomic weight 208.98; atomic number 83) does have a number of specialist applications. It is found in low concentrations in food, and daily dietary intake is in the region of 5–20 μg.[24] It can occur on occasion in large amounts due to accidental contamination. Its use in cosmetics and pharmaceuticals such as dusting powders, astringents, antacids and antiseptics make it a potential domestic contaminant of food. Because of its low melting-point and its ability to form alloys with other metals, bismuth is used in solders, fuses and safety devices. The presence of these in food-processing plants might result in product contamination.

Bismuth has not been shown to have any beneficial effects on man and it is not an essential nutrient. When injected into animals, bismuth compounds have caused liver and kidney damage.

Intake of large amounts of bismuth-containing pharmaceuticals has resulted in serious poisoning of children[25] and of some elderly patients.[26] There are no reports of poisoning due to ingestion of bismuth-containing food.

ZIRCONIUM

The human body contains about 0.4 g of zirconium (atomic weight 91.21; atomic number 40). It is found widely distributed in the lithosphere, with about 300 mg/kg in soils.[27] In fact, it is much more abundant than the more familiar lead, copper, nickel and zinc. The metal is hard and corrosion resistant and has a growing number of industrial applications. It is used, for example, as an alloy with iron to make a very tough steel. Its dioxide, zirconia, is of importance because of its excellent refractory qualities. Zircon, the naturally occurring silicate, is found in a variety of colours and is used as a semi-precious stone.

Zirconium occurs in most foods with, on average, about 1–3 mg/kg in fresh meat, but considerably less in most fruit and seafoods. Average daily intake has been calculated to be about 3.5 mg.[28]

Little is known about the absorption and metabolism of zirconium in the human body. In rats, zirconium compounds have a low toxicity when ingested. There is no indication that zirconium in the human diet is a cause for concern.

GERMANIUM

This is a metal which was once considered a rarity and of no industrial significance, but it has recently emerged as a metal of major importance. Germanium (atomic weight: 72.59; atomic number: 32) has the unique property of permitting an electric current to pass in only one direction. This rectifying power makes it invaluable in the electronic and allied industries for making transistors, which replace valves in the amplification of minute electric currents. The applications of transistors are manifold and in this way germanium is becoming increasingly used both domestically and industrially.

Germanium is widely distributed in food, but levels reported have generally been below 1 mg/kg.[29] Daily intake is probably about 0.4 mg.[30]

Animal experiments have indicated that absorption of germanium and its salts is rapid. Distribution to body tissues is uniform with, apparently, no accumulation in any particular organ. Excretion, mainly in urine, is rapid.

Ingestion of certain germanium compounds has been shown to produce toxic effects in experimental animals.[31] These included a reduction in life span, degenerative changes in liver and kidney and possible teratogenic effects. There is no evidence that similar effects occur in humans.

TUNGSTEN

Tungsten (atomic weight: 183.85; atomic number: 74) has the chemical symbol W, from its alternative name wolfram. It is a valuable, tough and hard metal, widely used in its own right as well as in steels and compounds. Electric-light filaments, contact points, cutting edges, machine tools and other high-speed machinery use tungsten and its alloys. Tungsten carbide is used as an abrasive. The metal finds its way, thus, into a variety of industries and situations where it can come into contact with food.

In spite of its wide use actual levels of tungsten in food and water have been shown to be low. Daily intake has been estimated at 8–13 μg.[32] Levels in domestic water in Sweden ranged from 0.03 to 0.1 μg/litre.[33]

Little is known about absorption of tungsten from food in man. The metal has not been shown to be an essential nutrient for man or animals. No reports of toxicity due to tungsten ingestion by humans have been reported, though some deleterious effects have been noted in animals fed on tungsten-containing feeds.[34]

One cause of concern with regard to tungsten ingestion is the fact that the metal has been shown to be an antagonist to the normal metabolic action of molybdenum. In experimental animals activities of xanthine oxidase and sulphite oxidase, both molybdenum-dependent enzymes, were reduced by tungsten.[35] Whether a similar antagonism occurs in humans is not known.

TELLURIUM

Tellurium (atomic weight: 127.6; atomic number 52) is a relatively rare metal of the sulphur family, very similar to selenium in chemical properties. It has a number of industrial uses, mainly in the metallurgical industry. Its presence improves the technological properties of steels and other alloys.

It is also used in the manufacture of glass, plastics and rubber. Small amounts are used as catalysts in the chemical industry, in explosives, anti-oxidants and also in various electronic devices. Some of its compounds have therapeutic uses. Tellurium occurs in foodstuffs at levels of 10–50 μg/kg, with daily intake about 100 μg.[36] About 10 per cent of ingested tellurium at most is absorbed. Accumulation occurs mainly in the skeleton. Total body content has been estimated to be about 8 mg in the adult. Excretion of tellurium in faeces and urine is rapid. A peculiar effect of tellurium ingestion is that a small amount is exhaled, apparently as dimethyltelluride. This has a garlic-like odour and accounts for the characteristic 'garlic breath' noted in animals and man following exposure to tellurium and its compounds.[37]

Serious metabolic effects have resulted from long-term exposure to tellurium compounds by experimental animals. These include liver and kidney damage as well as impairment of nervous function. Less obvious symptoms have been observed in humans exposed industrially to the element. Accidental ingestion of tellurium compounds in gram quantities resulted in a number of fatalities.[38] Recent investigations suggest that tellurium may have teratogenic effects in humans.[39]

THALLIUM

Thallium (atomic weight: 204.4; atomic number: 81) is widely distributed naturally, though in fairly small amounts. Most igneous rocks contain about 0.5 mg/kg and it is present at about 10 times that concentration in many soils. It is an extremely toxic metal and is used mainly as a pesticide, especially for rodents. Attempts have been made to restrict its use but, because of the emergence of warfarin-resistant rats, the use of the more effective thallium has actually increased in recent years. Thallium has a number of specialist uses in instruments and as a catalyst, but its use as a pesticide is of most significance from the point of view of food contamination.

Figures are not available for levels of thallium in normal food, but many cases of poisoning due to accidental contamination of food through contact with thallium-containing pesticides have been reported.[40]

Thallium is readily and probably totally absorbed from food.[41] It is widely distributed throughout the body, with particularly high accumulation in the kidneys. Excretion is in faeces, urine, saliva and hair. It is found in human breast milk and can cross the placental barrier.

Thallium compounds are cumulative in their poisonous effect. Acute poisoning usually results rapidly in nausea, diarrhoea and abdominal pain. Delirium, convulsions and circulatory involvement follow and culminate in death. When recovery occurs it may be complete, or mental abnormality may follow, as is often the case with children.[42]

Thallium poisoning through ingestion of barley containing thallium sulphate caused 27 cases of poisoning, with seven deaths. Others have resulted from the misuse of thallium-containing pharmaceuticals, such as ointments for treating ringworm. Of the 778 cases of accidental thallium poisoning reviewed in one report, 6 per cent were fatal. When poisonings and deaths due to industrial use of thallium are also considered, it is clear why an editorial in the *Lancet* called for the utmost vigilance in the maintenance of industrial hygiene where thallium was used, and recommended that the use of thallium-containing pesticides should be discontinued wherever possible and restrictions placed on their availability.[43]

THE REMAINING METALS

It is to be expected that several other metals besides those which have been discussed in these pages will find new or extended industrial and domestic uses in the future. As a result they may become food contaminants of note, deserving of at least some attention in a study such as this. A second edition might, for example, include indium and gallium and other metals not covered here, but which in the intervening years have attracted the attention of food scientists and toxicologists.

To some extent the increasing sophistication of analytical instruments and procedures will make this extension of knowledge inevitable. Determination at levels of accuracy and sensitivity quite beyond the analysts' powers of only a few years ago, methods which speed up and automate cumbersome sample preparations and, in addition, multi-element techniques which allow sample economy and speed, will contribute to the build-up of a store of toxicological and nutritional knowledge. This accumulation of data will result in a more refined appreciation of the role and significance of metals in the diet. It is certain that we will have to abandon in many cases our present relatively crude investigation of levels of total metals in food. We will be obliged to examine the particular form of chemical combination in which the metals occur. This will be connected with a need for added information on the bioavailability of elements. We will have to look more closely at the metals in the total context of the diet, at their interaction with other elements and, from the toxicological point of view, the influence of other factors, both physiological and psychological, on human dose response to metal contaminants in food.

As our scientific knowledge and our level of technological sophistication increase, so also we grow in our awareness that such advances are often accompanied by new problems or, possibly more often, by the detection of formerly unsuspected dangers. This is clearly the case with regard to the subject matter of this study. True, science and technology over the years have brought many benefits to mankind. Food is more abundant, its variety greater and the general nutritional status far better for many people than it was in the past. There is a growing level of awareness and an increased interest in the quality of life among ordinary people. Not least of their concerns is that related to the presence of contaminants in the food they eat. The food laws may be relied on in most cases to protect them against the dangers of such contamination. But, as we have seen, laws do not necessarily keep step with technological advances and often the food scientist or manufacturer will be faced with the problem of how to reconcile the benefits derived from the use of a particular substance with the adverse reaction of the public and his own concern at possible long-term risks to health of workers and consumers. It is hoped that this study will go some way towards helping those involved in the food-processing industry to make well-founded and reasonable judgements in such cases and that it will also help these and others concerned with food quality to have a realistic view of the complexity of the subject of metal contamination of food.

References

CHAPTER 1

1. MINISTRY OF AGRICULTURE, FISHERIES AND FOOD (1966). *Manual of Nutrition*, 8th edn (HMSO, London).
2. PEARSON, D. (1976). *The Chemical Analysis of Food*, 7th edn, 68–9 (Churchill Livingstone, Edinburgh).
3. NEBERGALL, W. H., SCHMIDT, F. C. and HOLTZSCLAW, H. F. (1968). *General Chemistry*, 3rd edn, 530 (Raytheon Educ. Co., Boston, Mass).
4. LISK, D. J. (1972). *Adv. Agron.*, **24**, 267–320.
5. PEIRSON, D. H. *et al.* (1973). *Nature (London)*, **241**, 252–6.
6. CRAUN, G. F. and MCCABE, L. J. (1975). *JAWWA*, **67**, 593–9.
7. SCHROEDER, H. A. (1971). *Am. J. Clin. Nutr.*, **24**, 562–73.
8. MURTHY, G. K. and RHEA, U. S. (1971). *J. Dairy Sci.*, **54**, 1001–5.
9. HOFFMAN, C. M. *et al.* (1968). *JAOAC*, **51**, 580–6.
10. SCHROEDER, H. A. and NASON, A. P. (1971). *Clin. Chem.*, **17**, 461–74.
11. SUMINO, K. *et al.* (1975). *Arch. Environ. Health*, **30**, 487–94.
12. PHIPPS, D. A. (1976). *Metals and Metabolism* (Clarendon Press, Oxford).

CHAPTER 2

1. DITTMER, H. J. (1937). *Am. J. Bot.*, **24**, 417–20, as quoted in E. Epstein, (1972). *Mineral Nutrition of Plants* (J. Wiley, New York).
2. HALL, N. S. *et al.* (1953). *North Carolina Agric. Exp. Sta. Tech. Bull.*, **101**, 1–40.
3. LISK, D. J. (1972). *Adv. Agron.*, **24**, 267–320.
4. REILLY, C. (1969). *New Phytol.*, **68**, 1081–7.
5. DUVIGNEAUD, P. and DENAEYER-DE SMET, S. (1963). *Bull. Soc. Roy. Bot. Belgique*, **96**, 93–231. Also WARREN, H. V. and DELAVAULT, R. E. (1949). *Bull. Geol. Soc. Am.*, **60**, 531.
6. REILLY, A. and REILLY, C. (1973). *New Phytol.*, **72**, 1041–6.

7. BRADSHAW, A. D., MCNEILLY, T. and GREGORY, R. P. G. (1965). *Industrialisation, Evolution and the Development of Heavy Metal Tolerance in Plants*, in *Ecology and Industrial Society*, ed. G. T. GOODMAN, R. W. EDWARDS and J. M. LAMBERT, *Brit. Ecol. Soc. Symp.*, **5**, 327.
8. REILLY, C. (1972). *Z. Pflanzenphysiol.*, **66**, 294–6.
9. REILLY, A. and REILLY, C. (1972). *Med. J. Zambia*, **6**, 125–7.
10. LITTLE, P. and MARTIN, M. H. (1972). *Environ. Pollut.*, **3**, 241–54.
11. MACKENZIE, E. J. and PURVES, D. (1975). *Chem. Ind.*, 4 Jan., 12–13.
12. BERROW, M. L. and WEBBER, J. (1972). *J. Sci. Food Agric.*, **23**, 93–100. Also WHEATLAND, A. B., SWANWICK, J. D. and TOMLINSON, E. J. (1976). *Water Pollut. Control*, **75**, 299–308.
13. PURVES, D. (1973). *Reclamation Industries International*, Sept./Oct., 17–21.
14. PIKE, E. R., GRAHAM. L. C. and FOGDEN, M. W. (1975). *JAPA*, **13**, 19–33.
15. STRENSTRÖM, T. and VAHTER, M. (1974). *Ambio*, **3**, 91–2.
16. WILLIAMS, C. H. and DAVID, D. J. (1973). *Aust. J. Soil Res.*, **11**, 43.
17. KOELLSTROM, T. *et al.* (1975). *Arch. Environ. Health*, **30**, 321–328.
18. JOHN, M. K. (1973). *Environ. Pollut.*, **4**, 7–15. Also SCHROEDER, H. A. and BALASSA, J. J. (1963). *Science*, **140**, 819.
19. CLARKSON, T. W. (1971). *Food Cosmet. Toxicol.*, **9**, 299–43.
20. ASSOCIATION OF PUBLIC ANALYSTS (1971). *Joint Survey of Pesticide Residues in Foodstuffs sold in England and Wales, 1 August 1967 to 31 July 1968*, London.
21. JERVIS, R. E. *et al.* (1970). *Mercury Residues in Canadian Foods, Fish and Wildlife* (University of Toronto, Toronto, Canada).
22. PRATER, B. E. (1975). *Water Pollut. Control*, **74**, 63–76.
23. WORKING PARTY ON MATERIALS CONSERVATION AND EFFLUENTS (1975). *Report of Industrial and Technical Committee of the Institute of Metal Finishing, Trans. Inst. Metal Finish.*, **53**, 197–202.
24. COLEMAN, A. K. (1975). *Chem. Ind.*, 5 July, 534–44.
25. BRENNAN, J. G. *et al.* (1976). *Food Engineering Operations*, 2nd edn (Applied Science Publishers, London).
26. GILFILLAN, S. C. (1965). *J. Occup. Med.*, **7**, 53–60.
27. BERITIC, T. and STAHULJAK, D. (1961). *Lancet*, **i**, 669.
28. BEECH, F. W. and CARR, K. G. (1977). *Cider and Perry*, in Rose, A. H. (ed.) *Economic Microbiology, Vol. I. Alcoholic Beverages*, 130–220. (Academic Press, London and New York).
29. HAMELLE, G. (1976). *Ann. Falsif. Expert Chim.*, **69**, 101–5.
30. HARPER, W. J. and HALL, C. W. (1976). *Dairy Technology and Engineering*, (Avi, Westport, Conn).
31. WHITMAN, W. E. (1975). *Food Progress*, **2**, 1–2.
32. SAMPAOLO, A. *et al.* (1971). *Rass. Chim.*, **23**, 226–33.
33. WHITMAN, W. E. (1978). *Proc. IFST*, **11**, 86–90.
34. PETRINO, P., CAS, M. and ESTIENNE, J. (1976). *Ann. Falsif. Expert. Chim.*, **69**, 87–99.
35. REILLY, C. (1978). *J. Food. Technol.*, **13**, 71–6.
36. REILLY, C. (1976). *HCIMA Rev.*, **2**, 34–40.
37. WALKER, A. R. P. and ARVIDSSON, V. B. (1950). *Trans. R. Soc. Trop. Med. Hyg.*, **47**, 536.
38. DROVER, D. P. and MADDOCKS, I. (1975). *PNG Med. J.*, **18**, 15–17.
39. KLEIN, M. *et al.* (1970). *N. Engl. J. Med.*, **283**, 669–72.

40. WATANABE, Y. *et al.* (1974). *Ann. Rep. Tokyo Metropol. Res. Lab. Pub. Health*, **25**, 293–6.
41. HEICHEL, G. H., HANKIN, L. and BOTSFORD, R. A. (1974). *J. Milk Food Technol.*, **37**, 499–503.
42. MERANGER, J. C., CUNNINGHAM, H. M. and GIROUX, A. (1974). *Can. J. Public Health*, **65**, 292–6.
43. KHANNA, S. K., SINGH, G. B. and HASAN, M. Z. (1976). *J. Sci. Food Agric.*, **27**, 170–4.
44. REILLY, C. (1973). *Ecol. Food. Nutr.*, **2**, 43–7.
45. SELBY, J. W. (1967). *Food Packaging, the Unintentional Additive*, in GOODWIN, R. W. L. (ed.), *Chemical Additives in Food*, 83–94 (Churchill, London).
46. VAN HAMEL ROOS (1891). *Rev. Intn. Falsif.*, **4**, 179, abstract in *Analyst* (1891) **16**, 195.
47. DRUMMOND, J. C. and MACARA, T. (1938). *Chem. Ind. (London)*, **57**, 828.
48. BECKHAM, I. BLANCHE, W. and STORACH, S. (1974). *Var Foeda*, **26**, 26–32.
49. PAGE, G. G., HUGHES, J. T. and WILSON, P. T. (1974). *Food Technol. N. Z.*, **9**, 32–5.
50. CROSBY, N. T. (1977). *Analyst*, **102**, 225–68.
51. CATALA, R. and DURAN, I. (1972). *Rev. Agroquim. Technol. Aliment.*, **12**, 319–28.
52. KIMURA, Y. *et al.* (1970). *Ann. Rep. Tokyo Metropol. Res. Lab. Pub. Health*, **22**, 95–100.
53. BOARD, P. W. (1973). *Food Technol. Aust.*, **25**, 15–16.
54. NEHRING, P. (1972). *Ind. Obst. Gemueseverwert*, **57**, 489–92.
55. BLUMENTHAL, A. and TROTTMAN, K. (1973). *Alimenta*, **12**, 141–4.
56. CHRZANOWSKI, S. and KRUPA, A. (1974). *Zesz. Cent. Lab. Premyslu Rybnego*, No 24, 73–80.
57. JIMENEZ, M. A. and KANE, E. H. (1974). *Compatibility of Aluminium for Food Packaging* (Am. Chem. Soc., Washington DC).
58. EHRLICH, P. R., EHRLICH, A. H. and HOLDEN, J. P. (1973). *Human Ecology Problems and Solutions*, 139 (Freeman, San Francisco).
59. MORRE, J. *et al.* (1977). *Bull. Acad. Vet. France*, **50**, 51–8.
60. AARKROG, A. and LIPPERT, J. (1969). *Environmental Radioactivity in Denmark in 1968* (Danish Atomic Energy Commission Research Establishment, Riso, Report No 201).
61. LUCAS, J. (1975). *Our Polluted Food*, 132 (C. Knight, London).
62. MITCHELL, N. T. (1968–72). *Technical Reports of the Fisheries Radiobiological Laboratories*, Lowestoft, FRL 2, 5, 7 and 8 (MAFF, London).
63. SIMPSON, R. E. BARATTA, E. J. and JELINEK, C. F. (1977). *J. Assoc. Off. Anal. Chem.*, **60**, 1364–8.

CHAPTER 3

1. CROSBY, N. T. (1977). *Analyst*, **102**, 225–68.
2. MAYRHOFER, J. (1891). *Chem. Zeit.*, **15**, 1054.
3. BODMER, R. and MOOR, C. G. (1897). *Analyst*, **22**, 141–3.
4. ALLEN, A. H. and COX, F. H. (1897). *Analyst*, **22**, 187–9.

5. WOLFF, J. (1898). *Ann. de Chimie Analyt.*, **2**, 105.
6. KAISER, H. (1899). *Chem. Zeit.*, **23**, 496.
7. KEBLER, L. F. and LA WALL, C. H. (1897). *Amer. Jour. Pharm.*, **69**, 244–50.
8. REUSS, W. (1891). *Chem. Zeit.*, **15**, 1522–3.
9. ACCUM, F. (1820). *Treatise on Adulterations of Food and Culinary Poisons and Methods of Detecting Them*, 2nd ed (Longmans, London; reissued by Mallinckodt, USA, 1966).
10. GILES, R. F. (1976). *The Development of Food Legislation in the U.K.*, in *Food Quality and Safety: A Century of Progress. Proceedings of the Symposium Celebrating the Centenary of the 'Sale of Food and Drugs Act 1875'*, 4, London October 1975 (MAFF/HMSO, London).
11. SCHMIDT, A. M. (1976). *Food and Drug Laws in the U.S.: A 200-year Perspective*, 151 ff., MAFF Symposium, October 1975, London.
12. HUTT, P. B. (1978). *Food Drug Cosmet Law J.*, **33**, 501–92.
13. TURNER, J. S. (1970). *The Chemical Feast* (Grossman, New York).
14. UNDERWOOD, E. J. (1971). *Trace Elements in Human and Animal Nutrition*, 3rd edn (Academic Press, London and New York).
15. LINDBERG, P. and LANNEK, N. (1970). *Trace Element Metabolism in Animals*, ed. Mills, C. F., 421–6 (Livingstone, Edinburgh).
16. MINISTRY OF AGRICULTURE, FISHERIES AND FOOD (1978). *Summary of Regulations and Recommendations for Heavy Metals for the United Kingdom*, Food Contamination Branch (MAFF, London).
17. PEARSON, D. (1976). *The Chemical Analysis of Food*, 7th edn, 70–2 (Churchill–Livingstone, Edinburgh).
18. LENANE, G. A. (1977). *Summary of Food Standards and Regulations*, in *Food Processing and Packaging Directory*, 16th edn, 499–646 (Consumer Industries Press, London).
19. NATIONAL RESEARCH COUNCIL COMMITTEE ON FOOD PROTECTION (1972). *Food Chemicals Codex*, 2nd edn, ix–xii (NAS, Washington).

CHAPTER 4

1. PEARSON, D. (1976). *The Chemical Analysis of Food*, 7th edn (Churchill–Livingstone, Edinburgh).
2. FISHER, R. A. and YATES, F. (1937). *Statistical Tables for Biological, Medical and Agricultural Research* (Oliver & Boyd, Edinburgh).
3. MERANGER, J. C. and SOMERS, E. (1968). *J. Assoc. Off. Anal. Chem.*, **51**, 922–4.
4. REILLY, C. (1973). *Ecol. Food Nutr.*, **2**, 43–7.
5. STRUNK, D. H. and ANDREASAN, A. A. (1967). *J. Assoc. Off. Anal. Chem.*, **50**, 388.
6. GORSUCH, T. T. (1959). *Analyst*, **84**, 135–73.
7. ANALYTICAL METHODS COMMITTEE (1960). *Analyst*, **85**, 643–56.
8. CROSBY, N. T. (1977). *Analyst*, **102**, 225–68.
9. AOAC (1975). *Official Methods of Analysis*, 12th edn, 461 (Association of Official Analytical Chemists, Washington, DC).
10. CORNER, M. (1959). *Analyst*, **84**, 41.

214 METAL CONTAMINATION OF FOOD

11. ANALYTICAL METHODS COMMITTEE (1959). *Analyst*, **84**, 214.
12. MIDDLETON, G. and STUCKEY, R. E. (1955). *Analyst*, **79**, 138.
13. ANALYTICAL METHODS COMMITTEE (1959). *Analyst*, **84**, 127.
14. BAETZ, R. A. and KENNER, C. T. (1975). *J. Agric. Food Chem.*, **23**, 41.
15. MACLEOD, A. J. (1973). *Instrumental Methods of Food Analysis* (Elek Science, London).
16. SANDELL, E. B. (1959). *Colorimetric Determination of Traces of Metals*, 3rd edn (Interscience, London).
17. HUNDLEY, H. K. and WARREN, E. C. (1970). *J. Assoc. Off. Anal. Chem.*, **53**, 705.
18. DELVES, H. T. (1976). *Essays Med. Biochem.*, **2**, 37–73.
19. DANKES, D. M. *et al.* (1973). *Science* **179**, 1140–2.
20. BROWNER, A. F. (1974). *Analyst*, **99**, 617–44.
21. CLEMENTE, G. F. (1976). *J. Radioanal. Chem.*, **32**, 25–41.
22. TANNER, J. T. and FRIEDMAN, M. H. (1976). *J. Radioanal. Chem.*, **37**, 529–38.
23. THOMAS, B. J. *et al.* (1976). *Symposium: Radioaktive Isotope in Klinik U. Forschung*, **12**, 155–64 (Bad Gastein, Germany).

CHAPTER 5

1. AGRICOLA, G. (1556). *De Re Metallica*, English translation by H. C. and L. H. Hoover (Dover Publications, New York, 1950).
2. DOWDING, M. F. (1978). *Met. Materials* (UK), July, 27–36.
3. NATIONAL ASSOCIATION OF RECYCLING INDUSTRIES (1978). *Booklet* (Nat. Assoc. Recycl. Ind., New York).
4. PERRIN, J. (1974) *Mater. Techniq.*, November, 518–20.
5. SCHROEDER, H. A. and NASON, A. P. (1971). Trace Element Analysis in Clinical Medicine, *Clin. Chem.*, **17**, 461–74.
6. BARRY, P. S. (1975). *Brit. J. Industr. Med.*, **32**, 119–39.
7. KEHOE, R. A. (1916). *J. Roy. Inst. Publ. Health Hyg.*, **24**, 181, 101–20, 129–43, 177–203. Also RABINOWITZ, M. B., WETHERHILL, G. W. and KOPPLE, J. D. (1974). *Environ. Health Perspect.*, **7**, 145.
8. ALEXANDER, F. W., DELVES, H. T. and CLAYTON, B. E. (1973). *Environmental Health Aspects of Lead*, 319–31 (Commission of European Communities).
9. HOTIUCHI, K. (1970). *Osaka City Med. J.*, **16**, 1–28.
10. MURTHY, G. K. and RHEA, U.S. (1971). *J. Dairy Sci.*, **54**, 1001–5.
11. TSUCHIYA, K. and SUGITA, M. (1971). *Nord. Hyg. Tidskr.*, **53**, 105–10.
12. TASK GROUP ON METAL ACCUMULATION (1972). *Accumulation of the toxic metals with special reference to their absorption, excretion and biological half-times: environmental physiology and biochemistry*, **3**, 65–107 (Proc. Int. Congr. Occup. Health).
13. TASK GROUP ON METAL TOXICITY (1976). *Effects and dose–response relationships of toxic metals*, ed. G. F. Nordberg, 1–111. (Elsevier, Amsterdam.)
14. SCHROEDER, H. A. and TIPTON, I. H. (1968). *Arch. Environ. Health*, **17**, 965–78.
15. WEISS, D., WHITTEN, B. and LEDDY, D., (1972). *Science*, **178**, 69.
16. PETERING, H. G., YEAGER, D. W. and WITHERUP, S. O. (1973). *Arch. Environ. Health*, **27**, 327–30.

17. REILLY, C. and HARRISON, F. (1979). *J. Hum. Nutr.*, **33**, 250–4.
18. NAS (1972). *Airborne Lead in Perspective* (National Academy of Science, National Research Council, Washington, DC).
19. REILLY, C. and REILLY, A. (1972). *Med. J. Zambia*, **6**, 125–7.
20. CHISOLM, J. J. (1973). *New Engl. J. Med.*, **289**, 1016–52.
21. WHO (1976). *Environmental Health Criteria 3: Lead* (WHO, Geneva).
22. INGLIS, J. A., HENDERSON, D. A. and EMMERSON, B. T. (1978). *J. Path.*, **124**, 65.
23. STOCKS, P. and DAVIES, R. I. (1960). *Br. J. Cancer*, **14**, 8–22.
24. IARC (1972). *Monographs on Evaluation of Carcinogenic Risk of Chemicals to Man*, Vol. 1, Intn. Agency for Research on Cancer, Lyon.
25. BINI, L. and BOLLEA, G. (1947). *J. Neuropathol.*, **6**, 271–8.
26. US DEPARTMENT OF HEALTH, EDUCATION AND WELFARE, PUBLIC HEALTH SERVICE, DIVISION OF AIR POLLUTION (1965). *Survey of Lead in the Atmosphere of Three Urban Communities*, Public Health Services Publication 999-AP-12 (Cincinnati Public Health Service).
27. SINGER, M. J. and HANSON, L. (1969). *Proc. Soil. Soc. Am.*, **33**, 152–3.
28. DAVIES, B. E. and HOLMES, P. L. (1972). *J. Agric. Sci. Camb.*, **79**, 479–84.
29. FAVRETTO, L., MARLETTA, G. P. and GABRIELLI, L. F. (1973). *At. Abs. Newsletter*, **12**, 101–3.
30. SCHMID, G., ROSOPULO, A. and WEIGELT, H. (1974). *Landwirtsch. Forsch. Sonderh.*, **31**, 150–9.
31. WILLIAMS, C. (1974). *J. Agric. Sci. Camb.*, **82**, 189–92.
32. NEWTON, G. D., SHEPHARD, W. W. and COLEMAN, M. S. (1974). *JWPCF*, **46**, 999–1000.
33. AUERMANN, E. and BÖRTITZ, S. (1977). *Nahrung*, **21**, 793–7.
34. KEHOE, R. A. (1961). *J. Roy. Inst. Pub. Health Hyg.*, **24**, 1–81, 101–20.
35. MINISTRY OF AGRICULTURE, FISHERIES AND FOOD (1972). *Survey of Lead in Food* (HMSO, London).
36. WHO (1976). *Environmental Health Criteria, 3. Lead* (WHO, Geneva).
37. NAS (1972). *Airborne Lead in Perspective* (National Academy of Sciences, National Research Council, Washington, DC).
38. CLARKSON, T. W. (1971). *Food Cosmet. Toxicol.*, **9**, 229–43.
39. LISK, D. J. (1972). *Adv. Agron.*, **24**, 267–320.
40. NATIONAL RESEARCH COUNCIL OF CANADA (1973). Lead in the Canadian Environment, NRCC No 13682 (Ottawa).
41. DEPARTMENT OF ENVIRONMENT (1977). *Pollution Paper No 12, Lead in Drinking Water, A Survey of Great Britain 1975–76* (HMSO, London).
42. CRAUN, G. F. and MCCABE, L. J. (1975). *JAWWA*, **67**, 593–9.
43. EGAN, H. (1972). *Lead in the Environment*, Hepple, P. (ed.), 34–42 (Institute of Petroleum, London).
44. BINNS, F., ENZOR, R. J. and MACPHERSON, A. L. (1978). *J. Sci. Food Agric.*, **29**, 71–4.
45. OH, S. J. (1975). *Arch. Phys. Med. Rehabil.*, **56**, 312–17.
46. KLEIN, M. *et al.* (1970). *New Engl. J. Med.*, **13**, 669–72.
47. WHITEHEAD, T. P. and PRIOR, A. P. (1960). *Lancet*, ii, 1343–4.
48. BERITIC, T. and STRAHULJAK, D. (1961). *Lancet*, i, 669.
49. HARRIS, R. W. and ELSEA, W. R. (1967). *J. Amer. Med. Assoc.*, **202**, 544–6.
50. WICHMANN, H. J. and CLIFFORD, P. A. (1935). *J. Assoc. Off. Anal. Chem.* **18**, 315.

51. PIERCE, J. O. *et al.* (1976). *The Determination of Lead in Blood; A Review and Critique of the State of the Art* (Internat. Lead Zinc Research Organization, New York).
52. YULE, H. P., LUKENS, H. R. and GUINN, V. P. (1965). *Nucl. Instr. Methods,* **33,** 277.

CHAPTER 6

1. HUGANIN, A. G. and BRADLEY, L. (1975). *J. Milk Food Technol.,* **38,** 285–300.
2. WHO (1976). *Environmental Health Criteria l. Mercury* (WHO, Geneva).
3. LANDNER, L. (1971). *Nature* (London), **230,** 452–4.
4. JENSEN, S. and JERNELOV, A. (1972). *Mercury Contamination in Man and His Environment* (International Atomic Energy Agency, Vienna).
5. SWEDISH EXPERT GROUP (1971). *Methyl Mercury in Fish. A Toxicological-Epidemiological Appraisal of Risks. Nord. Hyg. Tdskr.* Suppl. 4.
6. BOUQUIAUX, J. (1974). *CEC European Symposium on the Problems of Contamination of Man and His Environment by Mercury and Cadmium.* (CID Luxembourg).
7. BERLIN, M. (1972). *Technical Report Series No. 137* (International Atomic Energy Agency, Vienna).
8. SILLEN, L. G. (1963). *Sven Kem. Tidskr.,* **75,** 161.
9. MINISTRY OF AGRICULTURE, FISHERIES AND FOOD (1971). *Survey on Mercury in Food,* and (1973) *Supplementary Report on Mercury in Food* (HMSO, London).
10. MILLER, G. E. (1972). *Science,* **175,** 1121–2.
11. CLARKSON, T. W. *et al.* (1973). *Heavy Metals in the Aquatic Environment* (Vanderbilt University, Nashville).
12. DEPARTMENT OF THE ENVIRONMENT (1976). *Environmental Mercury in Man.* Pollution Paper No 10 (HMSO, London).
13. BERLIN, M., CARLSON, J. and NORSETH, T. (1975). *Arch. Environ. Health,* **30,** 307–13.
14. BERLIN, M. and ULLBERG, S. (1963). *Arch. Environ. Health,* **6,** 589–601, 602–9, 610–16.
15. AL-SHAHRISTANI, H. and SHIHAB, K. M. (1974). *Arch. Environ. Health,* **28,** 342–4.
16. BAKIR, F. *et al.* (1973). *Science,* **181,** 230–41.
17. OSLAND, R. (1970). *Spectrovision,* **24,** 11; BRAUN, R. and HUSBANDS, A. P. (1971). *Spectrovision,* **26,** 2; HATCH, R. W. and OTT, W. L. (1968). *Anal. Chem.,* **40,** 2085–7.
18. PYE UNICAM (1976). *Atomic Absorption Methods: Mercury in Canned Fish* (Cambridge, UK).
19. WESTÖÖ, G. (1966). *Acta. Chem. Scand.,* **20,** 2131.
20. TANNER, J. T. and FORBES, W. S. (1975). *Anal. Chim. Acta,* **74,** 17.
21. US ENVIRONMENTAL PROTECTION AGENCY (1975). *Scientific and Technical Assessment Report on Cadmium,* Star Series E.P.A. 600/6–75–003 (US Government Printing Office, Washington, DC).

22. TETWORTE, W. (1973). *Staub-Reinalt. Luft*, **33**, 422–31.
23. NATIONAL HEALTH AND MEDICAL RESEARCH COUNCIL (1978). *Report on Revised Standards for Metals in Food* (NHMRC, Canberra).
24. LISK, D. (1972). *Adv. Agron.*, **24**, 267–320.
25. MURTHY, G. K. and RHEA, U.S. (1971). *J. Dairy Sci.*, **54**, 1001–5.
26. HOFFMAN, C. M. *et al.* (1968). *JAOAC*, **51**, 580–6.
27. MINISTRY OF AGRICULTURE, FISHERIES AND FOOD (1973). *Working Party on the Monitoring of Foodstuffs for Heavy Metals, 4th Report, Survey of Cadmium in Food* (HMSO, London).
28. ELINDER, C. G. *et al.* (1976). *Arch. Environ. Health*, **31**, 292–302.
29. MAHAFFEY, K. R. *et al.* (1975). *Environ. Health Perspect.*, **12**, 63–9.
30. MINISTRY OF AGRICULTURE, FISHERIES AND FOOD (1978). *Summary of Metals in Food Regulations* (HMSO, London).
31. OSTERGAARD, K. (1977). *Acta Med. Scand.*, **202**, 193.
32. RAHOLA, T., AARAN, R. K. and MIETTINEN, J. K. (1972). *Assessment of Radioactive Contamination in Man*, 553–62, IAEA, Vienna International Atomic Energy Agency, Proceedings Series (Unipub., New York).
33. NORDBERG, G. F. (1972). *Environ. Phys. Biochem.*, **2**, 7–36.
34. FRIBERG, L. *et al.* (1974). *Cadmium in the Environment*, 2nd edn (CRC Press, Cleveland, Ohio).
35. TSUCHIYA, K., SEKI, Y. and SUGITA, M. (1972). *Organ and tissue cadmium concentrations of cadavers from accidental deaths* (Proc. 17th Intn. Cong. Occup. Health, Buenos Aires).
36. PISCATOR, M. (1972). *Cadmium toxicity—industrial and environmental experience* (Proc. 17th Intn. Cong. Occup. Health, Buenos Aires).
37. SCHROEDER, H. A. (1967). *Circulation*, **35**, 570–82.
38. SCHROEDER, H. A. (1976). *The Trace Elements in Nutrition* (Faber and Faber, London).
39. PISCATOR, M. and LARSSON S. E. (1972). *Retention and toxicity of cadmium in calcium deficient rats* (Proc. 17th Intn. Cong. Occup. Health, Buenos Aires).
40. INTERNATIONAL AGENCY FOR RESEARCH ON CANCER (1976). *IARC Monographs on the Evaluation of Carcinogenic Risk of Chemicals to Man, Vol. II, Cadmium, nickel, some epoxides, miscellaneous industrial chemicals and general considerations on volatile anaesthetics*, 39–74 (Lyon, France).
41. SHIRAISHI, Y., KURAHASHI, H. and YOSIDA, T. H. (1972). *Proc. Jpn. Acad.*, **48**, 133.
42. SCHROEDER, H. A. and MITCHENER, M. (1971). *Arch. Environ. Health*, **23**, 102–6.
43. ANALYTICAL METHODS SUBCOMMITTEE (1975). *Analyst*, **100**, 761–3.
44. SMITH, J. C., KENCH, J. E. and LANE, R. E. (1955). *Biochem. J.*, **61**, 698–701.
45. HOLAK, W. (1975). *JAOAC*, **58**, 777.
46. ANALYTICAL METHODS SUBCOMMITTEE (1969). *Analyst*, **94**, 1153.
47. HARVEY, T. C. *et al.* (1975). *Lancet*, **ii**, 1269–72.
48. KJELLSTRÖM, T. *et al.* (1975). *Recent Advances in the Assessment of the Health Effects of Environmental Pollution*, Vol. II, 2197–213 (Commission of the European Communities, Luxembourg).

CHAPTER 7

1. VALLEE, B. L., ULMER, D. D. and WACKERS, W.E.C. (1960). *Arch. Ind. Health*, **21**, 132–51.
2. PINTO, S. S. and MCGILL, C. M. (1953). *Ins. Med. Surg.*, **22**, 281–7.
3. FIERZ, U. (1965). *Dermatologica*, **131**, 41–58.
4. SCHROEDER, H. A. and BALASSA, J. J. (1966). *J. Chronic. Dis.*, **19**, 85–106.
5. LISK, D. J. (1972). *Adv. Agron.*, **24**, 267–320.
6. BOWEN, H. J. M. (1966). *Trace Elements in Biochemistry* (Academic Press, London and New York).
7. DRINKING WATER STANDARDS (Rev. 1962). *Public Health Service Pub. No 956* (US Govt Printers Washington, DC).
8. WICKSTROM, G. (1972). *Work Environ. Health*, **9**, 2–8.
9. BORGONO, J. M. and GREIBER, R. (1972). *Trace Substances in Environmental Health V*, Hemphill, D. (ed) 13–24 (Univ. of Missouri Press, Columbia).
10. YEH, S. (1963). *Nat. Cancer Inst. Monogr.*, **10**, 81–107.
11. Observer Newspaper (London), January 1979.
12. FRIBERG, L. (1978). *Internat. Conf. Heavy Metal in Environ.*, Toronto, *Symp. Proc.* **I**, 21–34 (Univ. of Toronto).
13. KELYNACK, T. N. *et al.* (1900). *Lancet*, **ii**, 1600–3.
14. TSUCHIYA, K. (1977). *Environ. Health Perspect.*, **19**, 35–42.
15. DORLE, M. and ZIEGLER, K. (1930). *Z. Klin. Med.*, **112**, 237–56.
16. CODE OF FEDERAL REGULATIONS. Title 21. Section 120. 192/3/5/6 and 133 g. 33.
17. HUNTER, F. T., KIP, A. F. and IRVINE, J. W. (1942). *J. Pharmacol Exp. Ther.*, **76**, 207–20.
18. VALLEE, B. L., ULMER, D. D. and WACKER, W. E. C. (1960). *Arch. Ind. Health*, **21**, 132–51.
19. LIEBSCHER, K. and SMITH, H. (1968). *Arch. Environ. Health*, **17**, 881–90.
20. CRECELIUS, E. A. (1977). *Environ. Health Perspect.* **19**, 147–50.
21. SMITH, H. (1964). *Forensic Sci. J.* **4**, 192–9.
22. LIEBSCHER, K. and SMITH, H. (1968). *Arch. Environ. Health.* **17**, 881–90.
23. LANDER, H., HODGE, P. R. and CRISP, C. S. (1965). *J. Forensic Med.*, **12**, 52–67.
24. NEUBAUER, O. (1947). *Brit. J. Cancer*, **1**, 192–251.
25. IARC Monographs (1973). *Evaluation of Carcinogenic Risk of Chemicals to Man. Some Inorganic and Organometallic Compounds*, Vol. 2 (International Agency for Research on Cancer, Lyon).
26. FERM, V. H., SAXON, A. and SMITH, B. M., (1971). *Arch. Environ. Health*, **22**, 557–60.
27. RHIAN, M. and MOXON, A. L. (1943). *J. Pharmacol. Exp. Ther.*, **78**, 249–64.
28. HOLMBERG, R. E. and FERM, V. H. (1969) *Arch. Environ. Health*, **18**, 873–7.
29. FROST, D. V. and SPRUTH, H. C. (1956). *Symposium on Medicated Feeds* (Medical Encyclopedia Inc., New York).
30. SANDELL, E. B. (1959). *Colorimetric Determination of Trace Metals*, 3rd ed (Inter-science Publishers, New York).
31. SMITH, K. E. and FRANK, C. W. (1968). *Appl. Spectrosc.* **22**, 765.
32. SMITH, H. (1959). *Anal. Chem.*, 1361–3.
33. BRAMAN, R. S. and FOREBACK, C. C. (1973). *Science*, **182**, 1247–9
34. DAUGHTREY, E. H., FITCHETT, A. W. and MUSHAK, P. (1975). *Anal. Chim. Acta*, **79**, 199–206.

35. MURTHY, G. K., RHEA, U. and PEELER, J. R. (1971). *Environ. Sci. Technol.* **5**, 436–42.
36. WESTER, P. O. (1974). *Atheroscler*, **20**, 207–15.
37. SCHMIDT, G. (1969). *Nuc. Sci. Ab.*, **23**, 3867.
38. SPENCER, D. W. *et al.* (1970). *J. Geoph. Res.*, **75**, 7688–96.
39. SOUKUP, A. V. (1972). *Trace Elements in Water*, in *Proceedings of Conference on Environmental Chemicals—Human and Animal Health* (Fort Collins, Colorado, August, 7–11).
40. BROWNING, E. (1969). *Toxicity of Industrial Metals* (Butterworths, London).
41. MONIER-WILLIAMS, G. W. (1934). *Antimony in Enamelled Hollow-Ware*, Ministry of Health Report on Public Health and Medical Subjects, No 7 (HMSO, London).
42. SUMINO, K. *et al.* (1975). *Arch. Environ. Health*, **30**, 487–94.
43. MANSOUR, M. M., RASSOUL, A. A. A. and SCHULERT, A. R. (1967). *Nature* (London), **214**, 819–20.
44. EL HALAWANI, A. A. (1968). *Bull. Endemic Dis.*, **10**, 123–33.
45. CHRISTIAN, G. D. and FELDMAN, F. J. (1970). *Atomic Absorption Spectrophotometry: Application in Agriculture, Biology and Medicine*, 421–3 (Wiley, New York).
46. THOMPSON, K. C. (1975). *Analyst*, **100**, 307–10.
47. TABOR, E. C. *et al.* (1970). *Health Lab. Sci.*, **7**, 92–5.
48. LISK, D. J. *Adv. Agron*, **24**, 267–320.
49. THORN, J. *et al.* (1978). *Br. J. Nutr.*, **39**, 391–6.
50. ROBINSON, M. F. (1976). *J. Human Nutr.*, **30**, 79–85.
51. SCHROEDER, H. A., FROST, D. V. and BALASSA, J. J. (1970). *J. Chron. Dis.*, **23**, 227–46.
52. THOMPSON, J. N., ERDODY, P. and SMITH, D. C. (1975). *J. Nutr.*, **105**, 274–82.
53. CRESTA, M. (1976). *Food and Nutr., (FAO)*, **2**, 8.
54. UNDERWOOD, E. J. (1971). *Trace Elements in Human and Animal Nutrition*, 3rd ed, 350 (London, Academic Press).
55. ARTHUR, D. (1972). *Can. Inst. Fd. Sci. Tech.*, **5**, 165.
56. EGAAS, E. and BRAEKKAN, O. R. (1977). *Fiskeridirektoratets Skrifter Serie Ernaering* **1** (3), 87–91.
57. FOOD AND NUTRITION BOARD (1976). *Nutr. Rev.* **34**, 347–50.
58. WHO (1973). *Technical Report Series No 532* (WHO, Geneva).
59. HALVERSON, A. W., GUSS, P. L. and OLSEN, O. E. (1963). *J. Nutr.*, **56**, 51.
60. MOXTON, A. L. (1938). *Science*, **88**, 81–3.
61. FRIEDEN, E. (1972). *Sci. Amer.*, **227**, 52–60.
62. NATIONAL HEALTH AND MEDICAL RESEARCH COUNCIL (1979). *Revised Standards for Metals in Food* (NH & MRC, Canberra).
63. SMITH, M. I., FRANKE, K. W. and WESTFALL, B. B. (1936). *Pub. Health Rep.*, **51**, 1496.

CHAPTER 8

1. BAUDART, G. A. (1975). *Rev. Alum.*, No. 438, 121–3.
2. SORENSON, J. R. J. *et al.* (1974). *Environ. Health Perspect.*, **8**, 3–95.

3. ZOOK, E. G. and LEHMANN, J. (1965). *JAOAC*, **48**, 850.
4. BEAL, G. D. *et al.* (1932). *Ind. Eng. Chem.*, **24**, 405.
5. SCHLETTWEIN-GSELL, D. and MOMMSEN-STRAUB, S. (1973). *Int. Z. Vitam. forsch*, **13**, 176–88.
6. CLARKSON, E. M. *et al.* (1972). *Clin. Sci.*, **43**, 519–31.
7. BERLYNE, G. M. *et al.* (1972). *Lancet*, **i**, 564–7.
8. GLAISTER, G. and ALLISON, A. (1913), *Lancet*, **i**, 843, Also ANON (1913). *Lancet*, **i**, 54. ANON (1928). *Lancet*, **ii**, 1246. BURN, J. H. (1932). *Aluminium and Food. A Critical Examination of the Evidence Available as to the Toxicity of Aluminium. British Non-Ferrous Metals Research Association, Research Reports, External Series No 162.*
9. SPIRA, L. (1933). *The Clinical Aspects of Chronic Poisoning by Aluminium and its Alloys* (Bale and Danielsson, London).
10. AMERICAN MEDICAL ASSOCIATION COUNCIL ON FOOD AND NUTRITION (1951). *J. Am. Med. Assoc.*, **146**, 477. Also US FOOD AND DRUG ADMINISTRATION (1971). *Safety of Cooking Utensils, Fact Sheet.*
11. LANDER, D. W., STEINER, R. L., ANDERSON, D. H. and DEHM, R. L. (1971). *Appl. Spectrosc.*, **25**, 270–5.
12. FERRIS, A. P. *et al.* (1970). *Analyst*, **95**, 574–8.
13. LANGMYHR, F. J. and TSALEV, D. L. (1977). *Anal. Chim. Acta*, **92**, 79–83.
14. FRITZE, K. and ROBERTSON, R. (1971). *J. Radional. Chem.*, **7**, 213–20.
15. KENT, N. L. (1942). *J. Soc. Chem. Ind.*, **61**, 183.
16. HAMILTON, E. I. *et al.* (1972). *Sci. Total Environ.*, **1**, 205–10.
17. TIPTON, I. H., STEWART, P. L. and MARTIN, P. G. (1966). *Health Phys.*, **12**, 1683–9.
18. MONIER-WILLIAMS, G. W. (1949). *Trace Elements in Food* (Chapman and Hall, London).
19. CARR, H. G. (1969). *Soc. Plast. Eng. Tech. Pap.*, **25**, 72–4.
20. SCHWARTZE, H. and CLARKE, J. (1927). *J. Pharmacol. Exp. Ther.*, **31**, 224.
21. HILES, R. A. (1974). *Toxicol. Appl. Pharmacol.*, **27**, 366–79.
22. DAVIDSON, S. *et al.* (1975). *Human Nutrition and Dietetics*, 5th edn, 140 (Churchill-Livingstone, Edinburgh).
23. BENOY, C. H., HOOPER, P. A., and SCHNEIDER, R. (1971). *Food Cosmet. Toxicol.*, **9**, 645–56.
24. CALLOWAY, D. H. and MCMULLEN, J. J. (1966). *Am. J. Clin. Nutr.*, **18**, 1–6.
25. SCHROEDER, H. A. *et al.* (1968). *J. Nutr.*, **96**, 37–45.
26. DE GROTT, A. P., FERON, V. J. and TIL, H. P. (1973). *Food Cosmet. Toxicol.*, **11**, 955–62.
27. ADCOCK, L. H. and HOPE, W. G. (1970). *Analyst*, **95**, 868–74.
28. JULIANO, P. O. and HARRISON, W. W. (1970). *Anal. Chem.*, **42**, 84–9.
29. BOWEN, H. J. M. (1972). *Analyst*, **97**, 1003–5.
30. RUDOLF, H., ALFREY, A. C. and SMYTHE, W. R. (1973). *Trans. Am. Soc. Artif. Intern. Organs*, **19**, 456–61.

CHAPTER 9

1. COTTON, F. A. and WILKINSON, G. (1966). *Advanced Inorganic Chemistry, A Comprehensive Text*, 625 ff. (Interscience Publishers, New York).

2. WHO (1973). *Technical Report Series No 532* (WHO, Geneva).
3. KLEVAY, L. M. (1975). *Nutr. Rep. Intn.*, **11**, 237–42.
4. ROBINSON, M. F. *et al.* (1973). *Br. J. Nutr.*, **30**, 195–205.
5. OWEN, C. A. (1964). *Amer. J. Physiol.*, **207**, 1203–6.
6. O'DELL, B. L. (1976). *Med. Clin. North. Am.*, **60**, 687–703.
7. SCHROEDER, H. A. *et al.* (1966). *J. Chronic. Dis.*, **19**, 1007–34.
8. REILLY, C. and HARRISON, F. (1979). *J. Hum. Nutr.*, **33**, 248–51.
9. STRICKLAND, G. T. *et al.* (1972). *Clin. Sci.*, **43**, 605–15.
10. LAHEY, M. E. and SCHUBERT, W. L. (1957). *Am. J. Dis. Childh.*, **93**, 31–4.
11. KARPEL, J. T. and PEDEN, W. H. (1972). *J. Pediat.*, **80**, 32–6.
12. PIMENTEL J. C. and MENEZES, A. P. (1975). *Am. Rev. Respir. Dis.*, **111**, 189–95.
13. CHUTTANI, H. K. *et al.* (1965). *Am. J. Med.*, **39**, 849–54.
14. HOLTZMAN, N. A. and HASLAM, R. H. A. (1968). *Pediat.*, **42**, 189–93.
15. SALMON, M. A. and WRIGHT, T. (1971). *Arch. Dis. Childh.* **46**, 108–10.
16. IVANOVICH, P., MANZLER, A. and DRAKE, R. (1969). *Trans. Am. Soc. Artif. Intern. Organs*, **15**, 316–20.
17. ANALYTICAL METHODS SUBCOMMITTEE, SOCIETY OF ANALYTICAL CHEMISTRY (1975) *Analyst*, **100**, 761–3.
18. MATOUSEK, J. P. and STEVENS, B. J. (1971). *Clin. Chem.*, **17**, 363–8.
19. DAVIDSON, S. *et al.* (1975). *Human Nutrition and Dietetics*, Sixth edn, 122 (Churchill Livingstone, Edinburgh).
20. WHO (1972). *Tech. Rep. Ser. No 503* (WHO, Geneva).
21. CROSBY, N. T. (1977). *Analyst*, **102**, 225–68.
22. SCHROEDER, H. A. (1971). *Am. J. Clin. Nutr.*, **24**, 562–73.
23. CRAUN, G. F. and McCABE L. J. (1975). *JAWWA*, **67**, 593–9.
24. BOTHWELL, T. H. and BRADLOW, B. A. (1960). *Arch. Path.*, **70**, 279–83.
25. LOWENTHAL, M. N. *et al.* (1967). *Med. J. Zambia*, **1**, 43–7.
26. DROVER, D. P. (1976). *Papua New Guinea Med. J.*, **19**, 111–12.
27. HAMBIDGE, K. M. (1978). *J. Hum. Nutr.*, **32**, 99–110.
28. MURTHY, G. K., RHEA, U. and PEELER, J. R. (1971). *Environ. Sci. Technol.*, **5**, 436–42.
29. LISK, D. J. (1972). *Adv. Agron.*, **24**, 267–320.
30. SCHROEDER, H. A. (1971). *Am. J. Clin. Nutr.*, **24**, 562–73.
31. BAETER, A. M. (1956). *Chromium* Vol. 1, Udy, M. J. (ed). *Am. Chem. Soc. Monograph* **132** (Reinhold, New York).
32. COMMITTEE ON BIOLOGICAL EFFECTS OF ATMOSPHERIC POLLUTANTS (1974). *Chromium* (Nat. Acad. Sci., Washington, DC).
33. McCABE, L. J. *et al.* (1970). *JAWWA*, **62**, 670–87.
34. CRAUN, G. F. and McCABE, L. J. (1975). *JAWWA*, **67**, 593–9.
35. BERROW, M. L. and WEBBER, J. (1972). *J. Sci. Food Agric.*, **23**, 93–100.
36. DONALDSON, R. M. and BARRERAS, R. F. (1966). *J. Lab. Clin. Med.*, **68**, 484–93.
37. HOPKINS, L. L. (1965). *Amer. J. Physiol.*, **209**, 731–5.
38. MERTZ, W. (1969). *Physiol. Rev.*, **49**, 163–239.
39. MERTZ, W. (1967). *Fed. Proc.*, **26**, 186–93.
40. MERTZ, W. *et al.* (1974). *Fed. Proc.*, **33**, 2275–80.
41. MERTZ, W. and CORNATZER, W. E. (1971). *Newer Trace Elements in Nutrition* (Dekker, New York).

42. MERTZ, W. (1975). *Nutr. Rev.*, **33**, 129–35.
43. TITUS, A. C. *et al.* (1930). *J. Ind. Hyg.*, **12**, 306–9.
44. BRIEGER, H. (1920). *J. Exp. Pathol. Ther.*, **21**, 393–408.
45. BEYERMAN, K. (1962). *Z. Anal. Chem.* **190**, 4–33, 346–69.
46. WILSON, L. (1968). *Anal. Chim. Acta*, **40**, 503–12.
47. WENLOCK, R. W., BUSS, D. H. and DIXON, E. J. (1979). *Br. J. Nutr.*, **41**, 253–61.
48. SCHROEDER, H. A. and NASON, A. P. (1971). *Clin. Chem.*, **17**, 461–74.
49. SCHROEDER, H. A. (1971). *Am. J. Clin. Nutr.*. **24**, 562–73.
50. SUMINO, K. *et al.* (1975). *Arch. Environ. Helath*, **30**, 487–94.
51. O'DELL, B. L., DE BOLAND, A. R. and KOIRTYOHANN, S. R. (1972). *J. Agric. Food Chem.*, **20**, 718.
    LEACH, R. M., MUENSTER, A. M. and WIEN, E. M. (1969). *Arch. Biochem. Biophys.*, **133**, 22.
52. VAN OETTINGEN, W. F. (1935). *Physiol. Revs.*, **15**, 175–201.
53. KIRKBRIGHT, G. D., SMITH, A. M. and WEST, T. S. (1966). *Analyst*, **91**, 700.
54. HARP, M. J. and SCOULER, F. I. (1952). *Br. J. Nutr.*, **47**, 67–72.
55. WESTER, P. O. (1974). *Atherosclerosis*, **20**, 207–15.
56. SCHROEDER, H. A. (1971). *Am. J. Clin. Nutr.*, **24**, 562–73.
57. NIX, J. and GOODWIN, T. (1970). *At. Absorpt. Newsl.*, **9**, 199–222.
58. STONE, I. (1965). *Wallerstein Lab. Commun.*, **28**, 209–17.
59. VALBERG, L. S., LUDWIG, J. and OLATUNBOSUN, D. (1979). *Gastroenterol.*, **56**, 241–51.
60. TAYLOR, D. M. (1962). *Phys. Med. Biol.*, **6**, 445–51.
61. HODGKIN, D. C. (1965). *Science*, **150**, 979–88.
62. SCHIRRMACHER, U. D. E. (1967). *Br. Med. J.*, **1**, 544–5.
63. MORIN, Y. and DANIEL, P. (1967). *Can. Med. Assoc. J.*, **97**, 926–8.
64. DAVIS, J. E. and FIELDS, J. P. (1955). *Fed. Proc.*, **14**, 331–2.
65. HUBBARD, D. M., CREECH, F. M. and CHOLAK, J. (1966). *Arch. Environ. Health*, **13**, 190–4.
66. SACHDEV, S. L., ROBINSON, J. W. and WEST, P. W. (1967). *Anal. Chim. Acta*, **38**, 499–506.
67. DELVES, H. T., SHEPHERD, G. and VINTER, P. (1971). *Analyst*, **96**, 260–73.
68. MITCHELL, R. L. (1945). *Soil Sci.*, **60**, 63.
69. SCHROEDER, H. A., BALASSA, J. J. and TIPTON, I. H. (1961). *J. Chronic Dis.*, **15**, 51–72.
70. LISK, D. (1972). *Adv. Agron.*, **24**, 267–320.
71. BULINSKI, R. and MICHNIEWSKI, J. (1977). *Bromstologia i Chemia Toksykologiczna* **10**, 215–16 (*Nutr. Abst. Revs. Series A*, (1978) **48**, 135).
72. SUNDERMAN, F. W. and SUNDERMAN, F. W., Jnr. (1961). *Amer. J. Clin. Path.*, **35**, 203–9.
73. CRAUN, G. F. and McCABE, L. J. (1975). *JAWWA*, **67**, 593–9.
74. BERROW, M. L. and WEBBER J. (1972). *J. Sci. Food Agric.*, **23**, 93–100.
75. DRINKER, K. R. *et al.* (1924). *J. Ind. Hyg.*, **6**, 307.
76. SUNDERMAN, F. W. (1965). *Amer J. Clin. Pathol.*, **44**, 182–200.
77. TIPTON, J. H. and COOK, M. J. (1963). *Health Phys.* **9**, 103.
78. SCHROEDER, H. A. and NASON, A. P. (1969). *J. Invest. Dermatol.*, **53**, 71–8.
79. DIXON, N. E. *et al.* (1975). *Jour. Amer. Chem. Soc.*, **97**, 4131–3.

80. NIELSON, T. H. (1974). *Trace Element Metabolism in Animals*, Vol. 2, 381–5. HOEKSTRA, W. G. *et al.* (eds) (University Park Press, Baltimore Md).
81. NOMOLO, B., MCNEELY, M. D. and SUNDERMAN, F. W. (1971). *Biochem*, **10**, 1647–51.
82. STOKINGER, H. E. (1963). *Nickel in Industrial Hygiene and Toxicology* Vol. II, Patty, F. A. (ed.) (Interscience, New York).
83. SUNDERMAN, F. W. *et al.* (1957). *Arch. Indust. Health*, **16**, 480–5.
84. SCHROEDER, H. A. (1970). *Arch. Environ. Health*, **21**, 798–806.
85. WESTERFIELD, W. W. and RICHERT, D. A. (1953). *J. Nutr.*, **51**, 85–90.
86. UNDERWOOD, E. J. (1971). *Trace Elements in Human and Animal Nutrition*. 3rd edn, 136 (Academic Press, New York and London).
87. BERROW, M. L. and WEBBER, J. (1972). *J. Sci. Food Agric.*, **23**, 93–100.
88. FERGUSON, W. S., LEWIS, A. H. and WATSON, S. J. (1943). *J. Agric. Sci.*, **33**, 44–51.
89. DOESTHALE, Y. G. and GOPALAN, C. (1974). *Brit. J. Nutr.*, **31**, 351–5.
90. YAROVAYA, G. A. (1964). *Sixth Int. Cong. Biochem.*, *New York*, **6**, 440.
91. KIRKBRIGHT, G. D., SMITH, A. M. and WEST, T. S. (1966). *Analyst*, **91**, 700.

CHAPTER 10

1. UNDERWOOD, E. J. (1971). *Trace Elements in Human and Animal Nutrition*, 3rd edn, 452–3 (Academic Press, London and New York).
2. SCHROEDER, H. A., BALASSA, J. J. and TIPTON, I. H. (1963). *J. Chron. Dis.*, **16**, 55–69.
3. LISK, D. J. (1972). *Adv. Agron.*, **24**, 267–320.
4. SCHROEDER, H. A. and NASON, A. P. (1971). *Clin. Chem.*, **17**, 461–74.
5. BERROW, M. L. and WEBBER, J. (1972). *J. Sci. Food Agric.*, **23**, 93–100.
6. LOWATER, F. and MURRAY, M. M. (1937). *Biochem J.*, **31**, 837.
7. CARLISLE, D. B. (1968). *Proc. Roy. Soc. B.*, **171**, 31–42.
8. GEYER, C. F. (1953). *J. Dent. Res.*, **32**, 590.
9. TIPTON, H. H., STEWART, P. L. and MARTIN, P. G. (1966). *Health Phys.*, **12**, 1683–9.
10. HARVEY, S. C. (1970). *The Pharmacological Basis of Therapeutics*, Goodman, L. S. and Gilman, A. (eds), 967–9 (Macmillan, New York).
11. KRONER, R. C. and KOPP, J. F. (1965). *JAWWA*, **57**, 150–6.
12. HEADLEE, A. J. W. and HUNTER, R. G. (1953). *Ind. Eng. Chem.*, **45**, 548–51.
13. HILL, W. B. and PILLSBURY, D. M. (1939). *Argyria, The Pharmacology of Silver*, (Williams, Baltimore).
14. MACINTYRE, D. *et al.* (1978). *Br. Med. J.*, **2**, 1749–50.
15. JENSEN, L. S. *et al.* (1974). *Poultry Sci.*, **53**, 57–64.

CHAPTER 11

1. REILLY, C. (1978). *Getting the Most out of Food*, Vol. 13, 47–69 (Van den Bergh and Jurgens, London).

2. PAUL, A. A. and SOUTHGATE, D. A. T. (1978). *McCance and Widdowson's The Composition of Foods*, 4th edn (London, HMSO).
3. BARRY, G. S. (1975). *CIM Bull. (Canada)*, **68**, 156–9.
4. ANON (1974). *Business Week* (USA) 14th December, 58–64.
5. MASIRONI, R., KOIRTYOHANN, S. R. and PIERCE, J. O. (1977). *Sci. Total Environ.*, **7**, 27–43.
6. MURPHY, E. W., WILLIS, B. W. and WATT, B. K. (1975). *J. Amer. Diet. Assoc.*, **66**, 345–5.
7. MINISTRY OF AGRICULTURE, FISHERIES AND FOOD (1976). *Manual of Nutrition*, 8th edn, (HMSO, London).
8. NATIONAL ACADEMY OF SCIENCES (1974). *Recommended Dietary Allowances*, 8th edn, (Washington, DC).
9. REITH, J. F., ENGELSMA, J. W. and VAN DITMARSH, W. C. (1976). *Voeding*, **37**, 498–507.
10. PRESTON, A. *et al.* (1972). *Environ. Pollut.*, **3**, 69–82.
11. CRAUN, G. F., and MCCABE, L. J. (1975). *JAWWA*, **67**, 593–9.
12. BERROW, M. L. and WEBBER, J. (1972). *J. Sci. Food Agric.*, **23**, 93–100.
13. SANSTEAD, H. H. (1973). *Am. J. Clin. Nutr.*, **26**, 1251–60.
14. REINHOLD, J. G. (1976). *Trace Elements in Human Health and Disease*, Prasad, A. S. and Oberleas, D. (eds), 163 (Academic Press, London and New York).
15. ECKHERT, C. D. *et al.* (1977). *Science*, **195**, 789–90.
16. HAHN, C. and EVANS, G. W. (1973). *Proc. Soc. Expl. Biol. Med.*, **144**, 793–5.
17. RICHARDS, M. P. and COUSINS, R. J. (1977). *Fed. Proc.*, **36**, 1106.
18. REILLY, C. and HARRISON, F. (1979). *J. Hum. Nutr.*, **33**, 250–4.
19. HAMBIDGE, K. M. (1978). *J. Hum. Nutr.*, **32**, 99–110.
20. BROWN, M. A. (1964). *Arch. Environ. Health*, **8**, 657–60.
21. REILLY, C. and MCGLASHAN, N. D. (1969). *S. Afr. J. Med. Sci.*, **34**, 43–8.

CHAPTER 12

1. PETZOW, C. T. and ZORN, P. (1974). *Chemiker–Zeit*, **98**, 236–41.
2. EISENBUD, M. *et al.* (1949). *J. Ind. Hyg. Toxicol*, **31**, 282–94.
3. WHITMAN, W. E. (1978). *Proc. IFST (UK)*, **11**, 86–90.
4. REEVES, A. L. (1965). *Arch. Environ. Health*, **11**, 209–14.
5. SCHUBERT, J. (1958). *Sci. Amer.*, **188**, 322–7.
6. SOHEL, A. E., GOLDFARB, A. R. and CRAMER, B. (1935) *J. Biol. Chem.*, **108**, 395–401.
7. RAMAKRISHNA, T. V., WEST, P. W. and ROBINSON, J. W. (1967). *Anal. Chim. Acta*, **39**, 81–7.
8. FRAME, G. M. *et al.* (1974). *Anal. Chem.*, **46**, 534–9.
9. WOLF, W. R. *et al.* (1972). *Anal. Chem.*, **44**, 616–18.
10. BOWEN, H. J. M. (1966). *Trace Elements in Biochemistry* (Academic Press, New York).
11. BOWEN, H. J. M. and DYMOND, J. A. (1955). *Proc. Roy. Soc.*, Ser. B., **144**, 355–76.
12. ALEXANDER, G. V., NUSBAUM, R. E. and MACDONALD, N. S. (1956). *J. Biol. Chem.*, **218**, 911–19.

13. COLE, V. V., HARNED, B. K. and HAFKESBRING, R. (1941). *J. Pharmacol. Exp. Ther.*, **71**, 1–5.
14. ADAMS, P. B. and PASSMORE, W. O. (1966). *Anal. Chem.*, **38**, 630.
15. BEESON, K. C. (1941). *US Dept. Agric. Misc. Pub.*, **369**, 120–1.
16. SCHROEDER, H. A., TIPTON, I. H. and NASON, A. P. (1972). *J. Chron. Dis.*, **25**, 491–517.
17. ROBINSON, W. O., WHETSTONE, R. R. and EDGINGTON, G. (1950). *US Dept. Agric. Tech. Bull.*, **1013**, 29–32.
18. BOSTRÖM, H. and WESTER, P. O. (1967). *Acta Med. Scand.*, **181**, 465–73.
19. CUDDIHY, R. G. and OZOG, J. A. (1973). *Health Phys.*, **25**, 219–24.
20. UNDERWOOD, E. J. (1977). *Trace Elements in Human and Animal Nutrition*, 3rd edn, 432 (Academic Press, London and New York).
21. ROZA, O. and BERMAN, L. B. (1971). *J. Pharmacol. Exp. Ther.*, **177**, 433–9.
22. KENT, N. L. and McCANCE, R. A. (1941). *Biochem. J.*, **35**, 837.
23. PFEIFFER, C. C., HALLMAN, L. F. and GERSH, I. (1945). *J. Amer. Med. Assoc.*, **128**, 266.
24. WOOLRICH, P. F. (1973). *Amer. Indust. Hyg. Assoc. J.*, **34**, 217–26.
25. BOYETTE, D. P. (1946). *J. Pediatr.*, **28**, 193–7.
26. ROBERTSON, J. F. (1974). *Med. J. Austral.*, **1**, 887–8.
27. LISK, D. (1972). *Adv. Agron.*, **24**, 267–320.
28. SCHROEDER, H. A. and BALASSA, J. J. (1966). *J. Chronic. Dis.*, **19**, 537.
29. SCHROEDER, H. A. and BALASSA, J. J. (1967a). *J. Chronic. Dis.*, **20**, 211–24.
30. HAMILTON, E. I. and MINSKI, M. J. (1972/73). *Sci. Tot. Environ.*, **1**, 375–94.
31. SCHROEDER, H. A. and BALASSA, J. J. (1967b). *J. Nutr.*, **92**, 245–53.
32. WESTER, P. O. (1974). *Atheroscler.*, **20**, 207–15.
33. BOSTROM, H. and WESTER, P. O. (1967). *Acta Med. Scand.*, **181**, 465–73.
34. KINARD, F. W. and VAN DE ERVE, J. (1941). *J. Pharmacol. Exp. Ther.*, **72**, 196–201.
35. COHEN, H. J. *et al.* (1973). *Proc. Natl. Acad. Sci.*, **70**, 3655–9.
36. NASON, A. P. and SCHROEDER, H. A. (1967). *J. Chron. Dis.*, **20**, 671–80.
37. HOLLINS, J. G. (1969). *Health Phys.*, **17**, 497–505.
38. KEAL, J. H. H., MARTIN, N. H. and TUNBRIDGE, R. E. (1946). *Br. J. Indust. Med.*, **3**, 175–6.
39. PATTON, G. R. and ALLISON, A. C. (1972). *Mutation Res.*, **16**, 332–6.
40. REED, D. *et al.* (1963). *J. Amer. Med. Assoc.*, **183**, 516–22.
41. LUND, A. (1956). *Acta Pharmacol. Toxicol.*, **12**, 251–68.
42. MUNCH, J. C., GINSBURG, H. M. and NIXON, C. E. (1933). *Jour. Amer. Med. Assoc.*, **100**, 1315–9.
43. ANON (1974). *Lancet*, **ii**, 564–5.

Index